CAMBRIDGE LIBRARY COLLECTION

Books of enduring scholarly value

Life Sciences

Until the nineteenth century, the various subjects now known as the life sciences were regarded either as arcane studies which had little impact on ordinary daily life, or as a genteel hobby for the leisured classes. The increasing academic rigour and systematisation brought to the study of botany, zoology and other disciplines, and their adoption in university curricula, are reflected in the books reissued in this series.

The Natural History of Birds

Georges-Louis Leclerc, Comte de Buffon (1707–88), was a French mathematician who was considered one of the leading naturalists of the Enlightenment. An acquaintance of Voltaire and other intellectuals, he work as Keeper at the Jardin du Roi from 1739, and this inspired him to research and publish a vast encyclopaedia and survey of natural history, the ground-breaking *Histoire Naturelle*, which he published in forty-four volumes between 1749 and 1804. These volumes, first published between 1770 and 1783 and translated into English in 1793, contain Buffon's survey and descriptions of birds from the *Histoire Naturelle*. Based on recorded observations of birds both in France and in other countries, these volumes provide detailed descriptions of various bird species, their habitats and behaviours and were the first publications to present a comprehensive account of eighteenth-century ornithology. Volume 7 covers woodpeckers and kingfishers.

Cambridge University Press has long been a pioneer in the reissuing of out-of-print titles from its own backlist, producing digital reprints of books that are still sought after by scholars and students but could not be reprinted economically using traditional technology. The Cambridge Library Collection extends this activity to a wider range of books which are still of importance to researchers and professionals, either for the source material they contain, or as landmarks in the history of their academic discipline.

Drawing from the world-renowned collections in the Cambridge University Library, and guided by the advice of experts in each subject area, Cambridge University Press is using state-of-the-art scanning machines in its own Printing House to capture the content of each book selected for inclusion. The files are processed to give a consistently clear, crisp image, and the books finished to the high quality standard for which the Press is recognised around the world. The latest print-on-demand technology ensures that the books will remain available indefinitely, and that orders for single or multiple copies can quickly be supplied.

The Cambridge Library Collection will bring back to life books of enduring scholarly value (including out-of-copyright works originally issued by other publishers) across a wide range of disciplines in the humanities and social sciences and in science and technology.

The Natural History of Birds

From the French of the Count de Buffon

VOLUME 7

COMTE DE BUFFON
WILLIAM SMELLIE

CAMBRIDGE UNIVERSITY PRESS

Cambridge, New York, Melbourne, Madrid, Cape Town, Singapore,
São Paolo, Delhi, Dubai, Tokyo, Mexico City

Published in the United States of America by Cambridge University Press, New York

www.cambridge.org
Information on this title: www.cambridge.org/9781108023047

This edition first published 1793
This digitally printed version 2010

ISBN 978-1-108-02304-7 Paperback

THE

NATURAL HISTORY

OF

BIRDS.

FROM THE FRENCH OF THE

COUNT DE BUFFON.

ILLUSTRATED WITH ENGRAVINGS;

AND A

PREFACE, NOTES, AND ADDITIONS,

BY THE TRANSLATOR.

IN NINE VOLUMES.

VOL. VII.

LONDON:

PRINTED FOR A. STRAHAN, AND T. CADELL IN THE STRAND
AND J. MURRAY, N° 32, FLEET-STREET.

M DCC XCIII.

CONTENTS

OF THE

SEVENTH VOLUME.

Page

THE Woodpeckers — — 1

The Green Woodpecker — — 6

FOREIGN BIRDS *of the Ancient Continent, which are related to the Green Woodpecker* 18

1. The Palalaca, or Great Green Woodpecker of the Philipines — ib.
2. Another Palalaca, or the Spotted Green Woodpecker of the Philipines 19
3. The Green Woodpecker of Goa 20
4. The Green Woodpecker of Bengal 21
5. The Goertan, or Green Woodpecker of Senegal — — 23
6. The Little Striped Woodpecker of Senegal 24
7. The Gray-headed Woodpecker of the Cape of Good Hope — — 25

BIRDS *of the New Continent, which are related to the Green Woodpecker* — 26

1. The Striped Woodpecker of St. Domingo ib.
2. The Little Olive Woodpecker of St. Domingo 28

 3. The

CONTENTS.

Page

3. The Green Striped Woodpecker of Cayenne 29
4. The Little Striped Woodpecker of Cayenne 30
5. The Yellow Woodpecker of Cayenne 32
6. The Ferruginous Woodpecker, *Lath.* 34
7. The Black breasted Woodpecker 35
8. The Rufous Woodpecker — 36
9. The Little Yellow-throated Woodpecker 37
10. The Least Woodpecker of Cayenne 38
11. The Gold-winged Woodpecker — 39
12. The Black Woodpecker — 41

BIRDS *of the New Continent, which are related to the Black Woodpecker* — — 46

1. The White-billed Woodpecker ib.
2. The Pileated Woodpecker — 48
3. The Lineated Woodpecker — 51
4. The Red-necked Woodpecker — 53
5. The Lesser Black Woodpecker 54
6. The Red-headed Woodpecker 55
7. The Greater Spotted Woodpecker 57
8. The Lesser Spotted Woodpecker 61

BIRDS *of the Ancient Continent, which are related to the Spotted Woodpecker* — 64

1. The Nubian Woodpecker, *Lath.* ib.
2. The Great Variegated Woodpecker of the Isle of Luçon — 65
3. The Little Brown-spotted Woodpecker of the Moluccas — — 66

BIRDS *of the New Continent, which are related to the Spotted Woodpecker* — — 67

1. The Spotted Woodpecker of Canada ib.
2. The Varied Woodpecker — 68
3. The Variegated Jamaica Woodpecker 69
4. The Striped Woodpecker of Louisiana 71

5. The

CONTENTS.

	Page
5. The Variegated Woodpecker of Encenada	72
6. The Hairy Woodpecker —	ib.
7. The Little Variegated Woodpecker of Virginia, *Buff.* — —	73
8. The Variegated Woodpecker of Carolina	74
9. The Variegated Undated Woodpecker	75
10. The Woodpecker Creepers —	77
The Wryneck — —	79
The Barbets — —	87
The Tamatia — —	88
The Tamatia with the Head and Throat Red	90
The Collared Tamatia — —	91
The Beautiful Tamatia — —	92
The Black and White Tamatias —	93
The Barbets — —	95
The Yellow-throated Barbet —	96
The Black-throated Barbet —	97
The Black-breasted Barbet —	98
The Little Barbet — —	99
The Great Barbet — —	100
The Green Barbet — —	101
The Toucans — —	103
The Toco — —	112
The Yellow-throated Toucan —	113
The Red-bellied Toucan —	116
The Cochicat — —	118
The Hotchicat — —	119
The Aracaris — —	120
The Grigri — —	ib.
The Koulick — —	122
The Black-billed Aracari —	124
The Blue Aracari — —	ib.
The Barbican — —	126
The Cassican — —	128

The

CONTENTS.

Page

The Calaos, or Rhinoceros Birds 130
The Tock — — 134
The Manilla Calao — — 137
The Calao, of the Island of Panay 138
The Molucca Calao — — 140
The Malabar Calao — — 142
The Brac, or African Calao — 147
The Abyssinian Calao — — 148
The Philippine Calao — — 150
The Round-helmeted Calao — 153
The Rhinoceros Calao — — 155
The King Fisher — — 158
Foreign King Fishers — — 173

Great KING FISHERS *of the Old Continent* 174

1. The Greatest King Fisher — ib.
2. The Blue and Rufous King Fisher 175
3. The Crab King Fisher — 176
4. The Thick-billed King Fisher 178
5. The Pied King Fisher — 179
6. The Crested King Fisher — 182
7. The Black-capped King Fisher 183
8. The Green-headed King Fisher 184
9. The King Fisher with Straw-coloured Head and Tail — — 185
10. The White-collared King Fisher 186

The Middle-sized KING FISHERS *of the Ancient Continent* — — 188

1. The Baboucard — — ib.
2. The Blue and Black King Fisher of Senegal — — 189
3. The Gray-headed King Fisher 190
4. Yellow-fronted King Fisher 191
5. The Long-shafted King Fisher 192

Small

CONTENTS.

Page

Small King Fishers *of the Ancient Continent* — — 193

1. The Blue-headed King Fisher ib.
2. The Rufous King Fisher — 194
3. The Purple King Fisher — ib.
4. The White-billed King Fisher 195
5. The Bengal King Fisher — 197
6. The Three-toed King Fisher 198
7. The Vintsi — — 199

The King Fishers *of the New Continent. Great Species* — — 201

1. The Taparara — — ib.
2. The Alatli — — 203
3. The Jaquacati — — 205
4. The Matuitui — — 207

Middle-sized King Fishers *of the New Continent* — — 209

1. The Green and Rufous King Fisher ib.
2. The Green and White King Fisher 210
3. The Gip-Gip — — 211

Small King Fishers *of the New Continent.*

1. The Green and Orange King Fisher 212
2. The Jacamars - — 213
3. The Jacamar properly so called 214
4. The Long-tailed Jacamar — 216
5. The Todies — — 218
6. The North American Tody — 219
7. The Tic-Tic, or South American Tody 221
8. The Orange-bellied Blue Tody 222

The Aquatic Birds — 224

The

CONTENTS.

Page

The Stork — — 243
The Black Stork — — 261

Foreign BIRDS *which are related to the Stork* 265
1. The Maguari — — ib.
2. The Couricaca — — 267
3. The Jabiru — — 270
4. The Nandapoa — — 275

The Crane 277
The Collared Crane — — 295

CRANES *of the New Continent* — 296
1. The White Crane — — ib.
2. The Brown Crane — — 299

Foreign BIRDS *which are related to the Crane* 301
1. The Numidian Crane — ib.
2. The Royal Bird — — 306
3. The Cariama — — 313
4. The Secretary, or Meſſenger 316
5. The Kamichi — — 323

The Common Heron — 329
The Great White Heron — 350
The Black Heron — 353
The Purple Heron — 354
The Violet Heron — 355
The White Garzette — ib.
The Little Egret — 357

HERONS *of the New Continent* — 361
1. The Great Egret — — ib.
2. The Rufous Egret — — 362
3. The Demi Egret — — 362
4. The Soco — — 364

5. The

CONTENTS.

	Page
5. The Black-capped White Heron	365
6. The Brown Heron —	ib.
7. The Agami Heron — —	366
8. The Hocti — —	367
9. The Houhou — —	368
10. The Great American Heron	369
11. The Hudson's Bay Heron —	370
The Crab Catchers — —	371
Crab Catchers of the Old Continent	372
1. The Squaiotta Heron —	ib.
2. The Rufous Heron —	373
3. The Chesnut Heron —	374
4. The Sguacco Heron —	375
5. The Mahon Crab Catcher —	376
6. The Coromandel Crab Catcher	ib.
7. The White and Brown Crab Catcher	377
8. The Black Crab Catcher —	ib.
9. The Little Crab Catcher —	378
10. The Blongios — —	379
Crab Catchers of the New World	381
1. The Blue Crab Catchers —	ib.
2. The Brown-necked Blue Crab Catcher	382
3. The Iron-gray Crab Catcher	383
4. The Red billed White Crab Catcher	384
5. The Cinereous Crab Catcher	385
6. The Purple Crab Catcher —	ib.
7. The Cracra — —	386
8. The Chalybeate Crab Catcher	387
9. The Green Crab Catcher —	388
10. The Spotted Green Crab Catcher	389
11. The Zilatat — —	390

CONTENTS.

	Page
12. The Rufous Crab Catcher with Green Head and Tail — —	390
13. The Gray Crab Catcher with Green Head and Tail — —	391
The Open Bill — —	392
The Bittern — —	394
Birds *of the Old Continent which are related to the Bittern* — —	405
1. The Greater Bittern —	ib.
2. The Little Bittern —	406
3. The Rayed Brown Bittern —	407
4. The Rufous Bittern —	408
5. The Little Senegal Bittern —	409
6. The Spotted Bittern —	410
Birds *of the New Continent which are related to the Bittern* —	411
1. The Starred Bittern —	ib.
2. The Yellow Bittern of Brazil	412
3. The Little Bittern of Cayenne	413
4. The Hudson's Bay Bittern —	414
5. The Onore — —	415
6. The Rayed Onore — —	416
7. The Onore of the Woods —	417
The Bihoreau — —	419
The Bihoreau of Cayenne —	422
The Tufted Umbre —	423
The Courliri, or *Courlan* —	425
The Savacou — —	426
The White Spoon Bill — —	431

The

CONTENTS.

	Page
The Woodcock — —	442
VARIETIES *of the Woodcock* —	458
1. The White Woodcock —	ib.
2. The Rufous Woodcock —	ib.
3. The Greater and Leſſer Woodcock	459
FOREIGN BIRD *which is related to the Woodcock*	460
1. The Savanna Woodcock —	ib.
The Snipe — —	463
The Jack Snipe — —	470
The Brunette — —	472
FOREIGN BIRDS *which are related to the Snipes*	473
1. The Cape Snipe — —	ib.
2. The Madagaſcar Snipe —	474
3. The China Snipe — —	475
The Barges —— ——	476
The Common Barge — —	479
The Barking Barge — —	480
The Variegated Barge —	481
The Rufous Barge — —	482
The Great Rufous Barge —	483
The Rufous Barge of Hudſon's Bay	484
The Brown Barge ——	485
The White Barge — —	486
The Horſemen — —	488
The Common Horſeman —	489
The Redſhank — —	490
The Striped Horſeman —	492
The Variegated Horſeman ——	494
The White Horſeman — —	496

The

CONTENTS.

Page
The Green Horseman — 497
The Ruff and Reeve — 498
The Maubeches — — 505
The Common Maubeche — *ib.*
The Spotted Maubeche — — 507
The Gray Maubeche — ib.
The Sanderling — — 508
The Green Sandpiper — 509
The Common Sandpiper — 514
The Sea Partridge — — 516
The Gray Sea Partridge — 517
The Brown Sea Partridge — 518
The Giarole — — ib.
The Collared Sea Partridge — 519
The Sea Lark — — 521
The Cincle — — 524

THE

THE

NATURAL HISTORY

OF

BIRDS.

The WOODPECKERS*.

Les Pics. Buff.
Pici. Linnæus, &c.

THOSE animals alone, which live upon the fruits of the earth, join in ſociety. Nature entertains them with a perpetual banquet, and abundance begets thoſe gentle peaceful diſpoſitions which are fitted for ſocial intercourſe. Other animals are conſtantly engaged in the purſuit of prey;

* In Greek the Woodpecker was called Δενδροκολαπτης, Δρυοκολαπτης from δενδρον, a tree, δρυς, an oak, and κολαπτω, to hollow out by blows; alſo Ξυλοκοπος, from ξυλον, wood, and κοπτω, to cut, and in Heſychius Σπελεκτος. Ariſtophanes calls it, Πελεκαν, from πελεκυς, an ax, alluding to the form and office of its bill: in modern Greek Κουρκουνι της. Pliny terms it *Picus Arborarius*. The Hebrew name is *Anapha*, and according to others, *Bleſchiat*. In Italian *Pico*, or *Picchio*. In German *Specht*. In Flemiſh *Spicht*. In Spaniſh *Bequebo*. In Poliſh *Dzieziol*. In Turkiſh *Sægarieck*.

 urged

urged by want, reſtrained by apprehenſions of danger, they depend for ſubſiſtence on the vigour of their own exertions: they have ſcarce time to ſatisfy their immediate deſires, and no leiſure to cheriſh the benevolent affections. Such is the ſolitary condition of all the carnivorous birds, except a few cowardly tribes which prowl on putrid carrion, and rather combine like robbers, than unite as friends.

And of all the birds which earn their ſubſiſtence by ſpoil, none leads a life ſo laborious and ſo painful as the Woodpecker. Nature has condemned it to inceſſant toil and ſlavery. While others freely employ their courage or addreſs, and either ſhoot on rapid wing, or lurk in cloſe ambuſh; the Woodpecker is conſtrained to drag out an inſipid exiſtence in boring the bark and hard fibres of trees, to extract its humble prey. Neceſſity never ſuffers any intermiſſion of its labours; never grants an interval of ſound repoſe: often during the night it ſleeps in the ſame painful poſture as in the fatigues of the day. It never ſhares the cheerful ſports of the other inhabitants of the air: it joins not their vocal concerts, and its wild cries, and ſaddening tones, while they diſturb the ſilence of the foreſt, expreſs conſtraint and effort. Its move-

movements are quick; its geſtures, full of inquietude; its looks, coarſe and vulgar; it ſhuns all ſociety, even that of its own kind; and when it is prompted by luſt to ſeek a companion, its appetite is not ſoftened by delicacy of feeling.

Such is the narrow and groſs inſtinct ſuited to a mean and a gloomy life. The organs with which the Woodpecker is furniſhed, correſpond to its deſtination. Four thick nervous toes, two turned forwards, and two backwards *, the one reſembling a ſpur, being longeſt and ſtouteſt; all of them armed with thick hooked nails, connected to a very ſhort and extremely muſcular foot, enable the bird to cling firmly, and to creep in all directions on the trunks of trees †. Its bill is edged, ſtraight, wedge-ſhaped, ſquare at the baſe, channelled longwiſe, flat, and cut vertical at its tip like a chiſel: this is the inſtrument with which it pierces the bark, and bores into the wood, to extract the inſects, or their eggs. The ſubſtance of the bill is hard and ſolid ‖, and riſes out of the cranium, which is very thick. Powerful muſcles act upon its ſhort neck, and direct its inceſſant blows, which

* Aldrovandus.

† Ariſtotle. *Lib. ix.* 9.

‖ Belon and Ariſtotle.

ſometimes penetrate even to the pith of the wood. It darts its long tongue, which is tapered and rounded like an earth-worm, and tipt with a hard bony point, like a needle. Its tail conſiſts of ten ſtiff quills, bent inwards, truncated at the ends, beſet with hard briſtles; and this often ſerves it as a reſt, while employed in a conſtrained, and often inverted, poſture. It breeds in the cavities which it has in part formed itſelf; the progeny iſſue from the heart of the tree, and, though furniſhed with wings, they are almoſt confined to the verge of its circumference, and condemned to tread the dull round of life.

The genus of the Woodpecker contains a great number of ſpecies, which differ in ſize and in colours. The largeſt is equal in bulk to the crow; and the ſmalleſt exceeds not the titmouſe. But few individuals are included in each ſpecies; which muſt ever be the caſe where a laborious courſe of life checks multiplication. Yet nature has placed Woodpeckers in all countries where ſhe has planted trees, and in greater plenty in the warm climates. There are only twelve ſpecies in Europe, and in the arctic regions; but we may reckon twenty-ſeven from the hot countries of America, of Africa, and of Aſia. And thus, though we we have conſiderably abridged the number, thirty-nine ſpecies ſtill remain; ſixteen of which were hitherto unknown.—We may obſerve in general, that the Woodpeckers of either continent differ from other birds in the ſhape of the feathers of the tail, which terminate in a point more or leſs ſharp.

The three ſpecies of Woodpeckers known in Europe are *the Green*, *the Black*, *and the Variegated*. Theſe have no varieties in our climates, and would ſeem to have migrated from the parent families in both continents. After theſe European ones, we ſhall range the foreign Woodpeckers akin to them.

The GREEN WOODPECKER*.

Le Pic Vert. Buff.
Picus Viridis. Linn. Gmel. &c. &c.
The Wood-Spite, Rain-fowl, High-hoe or *Hew-hole.* Will.

THIS is the beſt known, and the moſt common of the Woodpeckers. It arrives in the ſpring, and makes the foreſts reſound with the ſhrill harſh cries *tiacacan, tiacacan*, which are heard at a great diſtance. Theſe ſounds are uttered chiefly when it bounds in the air, ſinking, and again riſing by ſtarts, and deſcribing its waved tracks; but though it mounts only to a ſmall height, it can fly from wood to wood. In the pairing ſeaſon, it has, beſides its ordinary cry, a call of courtſhip, which reſembles in ſome meaſure a loud continued burſt of laughter *tio tio, tio, tio, tio,* repeated thirty or forty times in ſucceſſion †.

* Ariſtotle terms it Κολεὸς. Among the Romans it received the appellation *Picus Martius*, being conſigned to the God of War. In Italian it bears the names *Piccio*, and *Pico Verde*. In German, *Grun-Specht*. In Swediſh, *Wedknarr*, *Groen-joeling* and *Groen-Spick*. In Daniſh and Norwegian, *Gron-Spæt* and *Gnul-Spæt*. In Lapponic, *Zhiane*.

† Aldrovandus ſays, that it is ſilent in ſummer; probably it reſumes its cry in autumn, for in that ſeaſon we have heard it make the woods to reſound.

The

Nº 162

THE GREEN WOODPECKER.

The Green-Woodpecker is ſeen oftener on the ground || than the other Woodpeckers, particularly near ant-hills, where we may be ſure to find it, and even to catch it by means of a nooſe. It inſerts its long tongue into the narrow hole, through which the ants commonly iſſue, and as ſoon as it feels the tip covered with theſe inſects, it withdraws, and ſwallows them. But when theſe little republicans are inactive and ſtill, or torpid with cold, the bird aſſaults their citadel, and, employing both feet and bill, ſoon makes a breach, and at eaſe devours them, and their chryſalids.

At other times it creeps againſt the trees, which it ſtrikes with inceſſant blows; and, labouring with the moſt ſtrenuous activity, it often ſtrips them of all their bark. The ſtrokes of its bill are audible at a diſtance, and may be diſtinctly counted. In other reſpects it is indolent, and will ſuffer a perſon to approach near it, and will endeavour only to conceal itſelf from the ſportſman, by warping round the branch, and clinging on the oppoſite ſide. It has been ſaid, that, after a few knocks, it removes to another part of the tree to obſerve if it has pierced it; but its view is rather to

|| Willughby.

gather on the bark the insects which it has roused and put in motion: and, what is more certain, it judges from the sound of the blow in what cranies the worms are lodged, or where there is a proper cavity for its own accommodation.

It usually forms its nest in the heart of some worm-eaten tree, at the height of fifteen or twenty feet above the ground, and oftenest in the softer kinds of wood, the aspin, or the willow, rather than the oak. Both male and female, by turns, labour incessantly in boring the fresh part of the timber, until they penetrate to the rotten center. Then they fashion and enlarge the cavity, and throw out with their feet the chips and wood dust: sometimes they make the hole so deep and crooked, that the light cannot enter, and they rear their young in the dark. They commonly lay five eggs, which are greenish, with small black spots. The infant brood begin to creep before they are able to fly. The parents seldom leave them; they roost very early, and repose in their holes till day.

Some naturalists have thought the Green-Woodpecker was the rain-bird of the ancients, *pluviæ avis*, because it is generally believed to foretell rain by an unusual cry, which is drawling and plaintive, *pheu, pheu, pheu,*

pheu, and may be heard at a very confiderable diftance. Hence the Englifh call it *rain-fowl*, and fome provinces of France, fuch as Burgundy, it is vulgarly termed *the miller's procurer (procureur du meunier)*. Thefe obfervers allege, that as this bird fhews a forefight of the ftate of the atmofphere, fuperftition would naturally afcribe to it a more profound and wonderful fagacity. The Woodpecker held a principal rank among the aufpices *; its hiftory or fable was interwoven with the mythology of the ancient heroes of *Latium* †; its geftures were regarded as fignificant, and its appearance foreboded impending fate. Pliny relates a curious incident, which exhibits in the ancient Romans two qualities that might be deemed incompatible, fuperftitious obfervance, and elevation of fentiment ‡.

The

* *Pici Martii . . . in aufpicatu magni . . . principalis Latio funt in auguriis.* Plin. *Lib* x. 18.

† Picus, fon of Saturn, and father of Faunus, was grandfather of king Latinus. For defpifing the love of Circe, he was changed into a Green Woodpecker: he became one of the rural gods, under the name of *Picumnus*. While the fhe-wolf fuckled Romulus and Remus, this facred Woodpecker was feen to alight on their cradle. *See farther*, Gefner, p. 678.

‡ A woodpecker alighted on the head of the prætor Ælius Tubero, while he was fitting on his tribunal in the forum, and fuffered

The ſpecies of the Green Woodpecker is found in both continents; and, though it contains few individuals, they are widely ſpread. The Green Woodpecker of Louiſiana is the ſame with that of Europe ||; and that of the Antilles is only a variety §. Gmelin mentions his having ſeen, among the Tunguſe Tartars, a cinereous Green Woodpecker, which muſt be akin to the European *. Nor ſhall we heſitate to range with it the *gray-headed* Woodpecker from Norway, deſcribed by Edwards, and which

ſuffered itſelf to be taken by the hand. The augurs being conſulted on this prodigy, declared, that the empire was threatened with deſtruction, if the bird was liberated, and the prætor with death, if it was kept. Inſtantly Tubero tore it with his hands; ſhortly after, Pliny adds, the reſponſe was fulfilled. *Lib. x.* 18.

|| Dupratz.

§ There is a bird called *carpenter* in St. Domingo, no doubt, becauſe it ſhapes and hollows trees; if it be not the Green Woodpecker of Europe, it is a bird of the ſame ſpecies; it has its colours, its form, its note, and its habits. It does much injury to the palm trees, which it bores in many places, and often quite through, which makes them frail and periſhable. It is alſo very fond of the cocoa nut. We are obliged to hunt it when that fruit comes to maturity. *Note of the Chevalier Lefebvre Deſhayes.*

" * The Tunguſes of Nijaia-tunguoka aſcribe virtues to the cinereous Green Woodpecker; they roaſt this bird, pound it, mix it with any ſort of fat, except that of bears, becauſe this quickly grows rancid, and with this compound beſmear the arrows which they uſe in the chace: an animal ſtruck with one of theſe arrows, inſtantly falls. *Voyage en Siberie*, par Gmelin, tom. II. p. 113.

Klein

Klein and Brisson consider as a distinct species. In fact, the only difference between it and our Green Woodpecker, is, that its plumage is paler, and its head not marked with bright red, though there is a tint of that colour on the front. Edwards very justly attributes the alteration of its hues to the influence of climate. Brisson makes the *yellow Woodpecker* of Persia another * species, though in all probability it is nothing but a Green Woodpecker. Its size and almost its colours, are the same; and Aldrovandus formed his description from a figure exhibited at Venice; and such slender authority merits no attention.

Belon conceived the black Woodpecker to be a species of the Green Woodpecker; and this error has been adopted by Ray, who reckons two kinds of Green Woodpeckers. These oversights are occasioned by the misapplication of terms: such has also been the case with the appellation *picus martius*, which is often bestowed on the Woodpeckers

* Picus Persicus. *Gmel.*

Picus Luteus Persicus. *Briss.*

Picus Luteus Cyanopus Persicus. *Aldrov.*

Picchio Giallo (yellow). *Zinn.*

Specific character: it is yellow; its upper surface, the tips of the quills of its wings, and the spaces about its eyes are ferruginous.

in

in general, though originally it belonged exclusively to the Green Woodpecker.

Gesner has asserted, and Aldrovandus has endeavoured to prove, that the *Colios* of Aristotle was the Green Woodpecker: but almost all other naturalists have maintained that it was the Oriole. It may therefore be proper to discuss these opinions, both with a view to complete the history of these birds, and to elucidate two difficult passages in Aristotle.

In Theodore Gaza's edition of Aristotle, the word Κελεος, which he translates *Galgulus*, or Oriole, occurs twice in the same chapter. It is first represented as hostile to the Λιβυος, and then as associating with that bird, and haunting the sides of rivers and bushes *: that sort of life is not ascribed to the former, which to avoid confusion ought to be read Κολιος. And what Aristotle mentions in another passage †, when he treats more fully of the Κολιος ‡; that they are nearly as large as the turtle, that their voice is strong, &c. agrees perfectly with the character of the Green Woodpecker: but it has besides a

* Παρα ποταμον ϗ λοχμας; λοχμη denotes a thicket proper for an ambuscade.

† Lib. viii. 3.

‡ Observe, that he comprehends under the article birds that live upon insects and gnats.

property

property peculiar to itſelf, *that it eagerly bores the trees, and procures its ſubſiſtence from the rotten parts.* The epithet χλωρος too, which Ariſtotle beſtows, ſignifies *green*, not yellow, as Gaza renders it; and when we conſider that the antient philoſopher ranges the Κολιος after two Woodpeckers and before the creeper, we cannot for a moment doubt, but that he means by it the Green Woodpecker.

Albertus and Scaliger were aſſured that the Green Woodpecker could learn to ſpeak, and that it ſometimes pronounces words diſtinctly: Willughby juſtly diſbelieves it. The ſtructure of its tongue, which is long and worm-ſhaped, appears entirely unfit to articulate ſounds; and its wild intractable nature cannot ſubmit to education: for how could a bird be trained in the domeſtic ſtate, which lives only upon the inſects that lurk under the bark of trees*?

According

* The Viſcount de Querhoent aſſures us, however, that he kept them at leaſt for ſome time; but he confirms us in the idea of their untractable diſpoſition. I have ſeen, ſays he, young Green Woodpeckers which I was rearing, and which were ſtill in the neſt, fight with each other obſtinately. When I opened trees where was a brood, the parents conſtantly forſook them, and left them to periſh of hunger. The woodpeckers are vicious and quarrelſome; birds weaker than them are ever their victims; they break the ſcull with their bill, without afterwards preying on it. I had one in a room

According to Friſch, the males alone have red on the head; and Klein makes the ſame aſſertion. Salerne ſays, that they were miſtaken, and that the young ones have all the upper ſide of the head red, even in the neſt. According to the obſervation of Linnæus, this red varies, and appears mixed ſometimes with black ſpots, and ſometimes with grey ones, and, in a few inſtances, without any ſpots at all. Some individuals, which are probably old males, aſſume a red tint on the two black muſtachoes which ariſe from the corners of the bill, and their colours are in general more vivid.

Friſch relates, that in Germany during winter, the Green Woodpeckers plunder bee-hives. We doubt this fact, eſpecially as in France few or none of theſe birds remain through the inclement ſeaſon, and it is unlikely that the colder climate of Germany ſhould prove more inviting.

When we diſſect them, we commonly find the crop filled with ants. They have no *cæcum*, which is wanting equally in all birds of this kind; but inſtead of it there is

a room with partridges, and it killed them all one after another. When I entered, it climbed up my legs. It walked out into the fields, and returned to eat in its room. They are very familiar, but feel no attachment.

is a dilatation of the inteſtine. The gall-bladder is large; the alimentary canal two feet long; the right teſticle round, the left one oblong and arched, which is the natural ſtructure, ſince it was verified on a great number of ſubjects *.

But the mechaniſm of the tongue has been a ſubject of admiration among all naturaliſts. Borelli and Aldrovandus have deſcribed the form and functions of that organ: *Olaus Jacobæus*, in the Acts of Copenhagen, and Mery, in the Memoires of the Academy of Sciences at Paris, have delineated its curious anatomy. The tongue of the Green Woodpecker is, accurately ſpeaking, only the bony tip, and what is uſually taken for the tongue is the *os hyöïdes* inveſted with a membranous coat, and extending backwards into two long branches, at firſt oſſeous, and afterwards cartilaginous: theſe, after encircling the windpipe, reflect towards the head, and running cloſe in a furrow along the ſkull, they are inſerted on the forehead at the root of the bill. They are elaſtic cords, furniſhed with an apparatus of muſcles, both *extenſors* and *retractors*, which ſerve to move and direct this ſort of tongue. The whole is ſheathed

* Willughby.

by

by the prolongation of the ſkin, which lines the lower mandible, and which extends when the *os hyöides* is protruded, and collapſes, in annular wrinkles, as that bone is retracted. The bony tip, which is the real tongue, is connected to the extremity, and covered with a ſcaly horn, beſet with ſmall hooks bent back: and that it may be capable both to hold and to pierce its prey, it is naturally moiſtened with a viſcous fluid, that diſtills from two excretory ducts which riſe from a double gland.—After this ſtructure the tongue of all the Woodpeckers is faſhioned; indeed we might conclude from analogy, that it alſo obtains in ſuch birds in general, as protrude their tongue by extending it.

The Green Woodpecker has a very large head, and can briſtle the red feathers that cover its crown, which induced Pliny to term it tufted *. It is ſometimes caught by the decoy, but very rarely: it anſwers not the call ſo much as the noiſe made by ſtriking the tree where it lodges, and which reſembles that uſually occaſioned by its own boring. Sometimes it is ſeized by the neck in ſprings, as it creeps along the ſtake. But it is very coarſe food, and always ex-

* *Cirrhos pico martio.*

ceedingly

ceedingly lean and dry; though Aldrovandus ſays, that theſe birds are eaten in winter at Bologna, and are then pretty fat: this acquaints us, at leaſt, that they remain during that ſeaſon in Italy, while they diſappear in France [A].

[A] Specific character of the *Picus Viridis*; it is green, its head crimſon.

FOREIGN BIRDS
OF THE
Ancient Continent, which are related to the
GREEN WOODPECKER.

The PALALACA,
OR
GREAT GREEN WOODPECKER of the Philippines.

First Species.

Picus Philipparum. Lath.

CAMEL, in his account of the birds that inhabit the Philippines, and Gemelli Carreri, agree, that in thoſe iſlands there is a ſpecies of Green Woodpecker as large as an ordinary hen; meaning probably with regard to length, and not to bulk. It is called *Palalaca* by the iſlanders, but *Herrero*, or the Forger, by Spaniards, on account of the loud noiſe which it makes in ſtriking againſt the trees, and which may be heard, ſays Camel, at the diſtance of three hundred paces. Its voice is coarſe and raucous; its head red and tufted; its plumage of a green ground. Its bill is extremely firm and ſolid, and enables it to excavate its neſt in the hardeſt trees.

Another PALALACA,

OR THE

SPOTTED GREEN WOODPECKER of the Philippines.

Second Species.

Picus Manillensis. Gmel.

The Manilla Green Woodpecker. Lath

THIS differs entirely from the former in its size and colours. Sonnerat calls it the *Speckled Woodpecker (Pic Grivelé)*. It is of an intermediate bulk between the variegated and Green Woodpeckers, though nearer that of the latter: on each feather in the whole of the foreside of the body, there is a spot of dusky white, framed in blackish brown, which forms a rich enamel; the mantle of the wings is rufous, tinged with aurora-yellow, which, on the back, assumes a more brilliant hue, verging on red; the rump is carmine; its tail rusty gray; and its head bears a tuft, waved with yellowish rufous on a brown ground.

The GREEN WOODPECKER of Goa.

Third Species.

Picus Goensis. Gmel.

It is ſmaller than the European. The red feathers on its head are gathered into a tuft, and its temples are bordered by a white ſtripe, which widens on the arch of the neck; a black zone deſcends from the eye, and tracing a zigzag, falls upon the wing, whoſe ſmall coverts are equally black; a fine gold ſpot covers the reſt of the wing, and terminates in greeniſh yellow on the ſmall quills; the great ones are as it were indented with ſpots of greeniſh white, on a black ground; the tail is black; the belly, the breaſt, and the foreſide of the neck, as far as under the bill, are mailed lightly with white and black. This bird is one of the moſt beautiful of the Woodpeckers; it bears a ſtriking reſemblance to the following, which, joined to the circumſtance, that they inhabit contiguouſly, would induce us to conclude that they are the ſame, or at leaſt two kindred ſpecies.

The GREEN WOODPECKER of Bengal.

Fourth Species.

Picus Bengalensis. Linn. Gmel. Klein and Gerini.
The Bengal Creeper. Albin.
The Spotted Indian Woodpecker. Edw.

IT is of the ſame ſize with the preceding, and ſimilar to it. The gold colour is more ſpread on the wings, and covers the back alſo; a white line, riſing from the eye, deſcends on the ſide of the neck like the black zigzag of the Goa Woodpecker; the tuft, though more diſplayed, appears only on the back of the head *, whoſe crown and foreſide are clothed with ſmall black feathers, beautifully ſpotted with white drops; the plumage under the bill, and on the throat, is the ſame in both birds; the breaſt and ſtomach are white, croſſed and mailed with blackiſh and brown, but leſs ſo in this than the preceding. Theſe minute differences would not perhaps be ſufficient to diſtinguiſh the two ſpecies; but

* A character more remarkable than that of *black nape*, by which Linnæus defines the ſpecies.

 the

the Goa Woodpecker has its bill one third longer than that of Bengal [A].

With the Bengal Woodpecker, I ſhall range not only the Green Bengal Woodpecker of Briſſon †, but alſo his Cape-of Good-Hope Woodpecker ‡, which indeed bears a cloſer reſemblance. The reaſon perhaps is, that the one from the Cape-of Good-Hope was deſcribed from nature, while the other was taken from Edward's figure, which is only ſomewhat larger than our Green Bengal Woodpecker. Albin, who deſcribes the ſame bird, repreſents it as ſtill larger, and as equal in bulk to the European. But notwithſtanding theſe differences in the colours and ſize, it is eaſy to ſee the ſame bird through the three deſcriptions.

[A] Specific character of the *Picus Bengalenſis*: it is green, its creſt red, its nape black; below and before it is white ſpotted with black.

† *Picus Viridis Bengalenſis*. Briſſ.

Thus deſcribed: " it is creſted; above yellowiſh green, below white; the margins of its feathers, brown; its creſt, red; the fore part of its head, and the lower part of its neck, variegated with white and black; the upper part of its neck, black; a bright white bar extending from the eyes along the ſides of the neck; its tail quills blackiſh, ſhaded with dull green."

‡ *Picus Capitis Bonæ Spei*. Briſſ.

Thus deſcribed: " it is orange above, ſhining with a golden hue; below dirty white; the margins of its feathers, brown; the upper and back part of its head, red; the upper part of its neck and its rump, blackiſh; a bright white bar extended from the noſtrils below the eyes, and along the ſides of the neck; its tail-quills, blackiſh."

The GOERTAN,

OR

Green Woodpecker of Senegal.

Fifth Species.

Picus-Goertan, Gmel.

The Crimson-rumped Woodpecker.

THIS Woodpecker, which is termed *Göertan* at Senegal, is not so large as the Green Woodpecker, and scarcely equal to the variegated one. The upper surface of the body is brown-gray, tinged with dull greenish, spotted on the wings with waves of faint white, and interrupted on the head and rump by two marks of fine red; all the under surface of the body is gray stained with yellowish. This species and the two following were unknown to naturalists.

The Little Striped WOODPECKER of Senegal.

Sixth Species.

Picus Senegalenſis. Gmel.

The Gold-backed Woodpecker. Lath.

THIS Woodpecker is not larger than a ſparrow; the upper ſide of its head is red; a brown half-maſk paſſes over the front, and behind the eye; the plumage, which is waved on the fore part of the body, exhibits ſmall feſtoons, alternately brown, gray, and dull white; the back is of a fine gold fulvous, which alſo tinges the great quills of the wing, whoſe coverts, as well as thoſe of the rump, are greeniſh. Though much inferior in ſize to the European Woodpeckers, we ſhall find that this African ſpecies is by no means the ſmalleſt of this extenſive genus.

The Gray-headed WOODPECKER of the Cape of Good Hope.

Seventh Species.

Picus Aurantius. Linn. and Gmel.

Picus Capitis Bonæ Spei. Briſſ.

The Orange Woodpecker. Lath.

ALMOST all the Woodpeckers have a mottled plumage, but in the preſent no colours are ſet in contraſt. A dull olive brown covers the back, the neck and the breaſt; the reſt of the plumage is deep gray, which is only ſomewhat lighter on the head; there is a red tinge at the origin of the tail.—This Woodpecker is not ſo large as a lark. [A]

[A] Specific character of the *Picus Aurantius:* above it is orange; its nape, its rump, and the quills of its tail are black.

BIRDS of the New Continent,

Which are related to the

GREEN WOODPECKER.

The Striped WOODPECKER of St. Domingo.

First Species.

Picus Striatus. Gmel.

Picus Dominicensis Striatus. Briss.

The Rayed Woodpecker. Lath.

BRISSON describes this bird in two different places; first under the name of *the Striated Woodpecker of St. Domingo*, and again under that of *the Little Striated Woodpecker of St. Domingo*, which he asserts to be smaller than the former, though the measures which he assigns in detail are the same; and with the salvo, *that the second may be the female of the first*, he regards them as two distinct species. But a single inspection of the figures will suffice to shew, that the differences result solely from age

or

or ſex. In the firſt, the crown of the head is black; the throat gray; the olive tinge of the body lighter, and the black ſtripes on the back are not ſo broad, as in the ſecond, which has the whole of the crown of the head red, and the fore part of the body pretty dull, with the throat white; but, in other reſpects, their ſhape and plumage are perfectly ſimilar. This bird is nearly as large as the variegated Woodpecker; all its upper garb is cut tranſverſely with black and olive bars; the green tinge appears on the gray of the belly, and more vividly on the rump, whoſe extremity is red; the tail is black.

The Little Olive WOODPECKER of St. Domingo.

Second Species.

Picus Passerinus. Linn. and Gmel.
Picus Dominicensis Minor. Briss.
The Passerine Woodpecker. Lath.

THIS species is six inches long, and nearly of the size with the lark; the crown of the head red, and its sides rusty gray; all the upper surface is yellowish olive, and all the under surface striped across with whitish and brown; the quills of the wing are olive, like the back, on the outside, and on the inside brown, and fringed on the edge with whitish spots, deeply engrained; a character in which it resembles also the Green Woodpecker: the feathers of the tail are gray mixed with brown. Notwithstanding its diminutive size, this Woodpecker is one of the stoutest; and it pierces the hardest trees. It is alluded to in the following extract from the History of the Buccaneers: " The carpenter is a bird not larger than a lark; its bill is about an inch

inch long, and ſo ſtrong, that in the ſpace of one day it will bore into the heart of a palm-tree: and we may obſerve, that this wood is ſo hard, as to ſpoil the edge of our beſt cutting tools [A]."

[A] Specific character of the *Picus Paſſerinus:* it is yellowiſh-olive, ſtriped below with brown and bright whitiſh.

The Great Striped WOODPECKER of Cayenne.

Third Species.

Picus Melanochloros. Gmel.

The Gold-creſted Woodpecker. Lath.

WE make no doubt but that this is the ſame with the *American creſted variegated Woodpecker* *, deſcribed incompletely by Briſſon, from a paſſage of Geſner. The creſt is of a gold fulvous, or rather aurora-red; there is a purple ſpot at the corner of the bill; the feathers are fulvous and black, with which the whole body is alternately

* *Picus varius Americanus Criſtatus.* Briſſ.

Thus deſcribed: " it is creſted, variegated with fulvous and black; its creſt gold fulvous, its cheeks reddiſh; a purple ſpot between its bill and its eyes; its tail-quills black."

variegated:

variegated: and these characters are sufficient to discriminate it. It is of the same size with the Green Woodpecker; its plumage is richly mailed with yellowish fulvous and fine black, which intermingle in waves, in spots, and in festoons; a white space in which the eye is placed, and a black tuft on the front, give a marked aspect to this bird, and which is still heightened by the red crest and purple mustachio [A].

[A] Specific character of the *Hirundo Melanochloros*: this is variegated with black and bright yellow, its crest golden, its tail black.

The Little Striped WOODPECKER of Cayenne.

Fourth Species.

Picus Cayanensis. Gmel.

The Cayenne Woodpecker. Lath.

Of the Striped Woodpeckers, which Brisson ranges after the variegated Woodpecker, several belong undoubtedly to the Green Woodpecker. This is particularly true of the Striped Woodpeckers of St. Domingo, and that of Cayenne, which we are now to describe. In fact, these three have a yela yellowiſh green caſt, analogous to the colour of the Green Woodpecker, and the undulated rays that ſpread on the plumage ſeem to be enlarged from the model of thoſe which mark the wing of the European bird.

The Little Striped Woodpecker of Cayenne is ſeven inches and five lines in length. It reſembles much the Striped Woodpecker of St. Domingo in its colours, but is ſmaller; black waved bars extend on the olive gray brown of its plumage; gray, fringed with black, covers the two exterior quills of the tail on each ſide, the ſix others are black; the back of the head is red; the front and throat are black, only this black is interſected by a white ſpot lying under and extending back.

The Yellow WOODPECKER of Cayenne.

Fifth Species.

Picus Exalbidus. Gmel.
Picus Cayanenſis Albus. Briſſ.
Picus Flavicans. Lath.

THOSE birds, which are enamoured of the ſolitude of the deſert, have multiplied in the vaſt foreſts of the new world, and the more ſo, as there man has yet encroached little on the antient domains of nature. We have received ten ſpecies of Woodpeckers from Guiana, and the Yellow Woodpeckers ſeem peculiar to that country. Moſt of theſe are ſcarcely known to naturaliſts, and Barrere has only noticed a few. The firſt ſpecies is deſcribed by Briſſon under the name of *White Woodpecker* *: its plumage is of a ſoft yellow; the tail black; the great quills of the wing brown, and the middle ones rufous; the coverts of the

* *Picus Cayanenſis Albus.* Briſſ.
Thus deſcribed: " it is dirty-white; a red longitudinal bar on either ſide upon the lower jaw; its tail-quills blackiſh."

wings

wings are brown gray, fringed with yellowiſh white. It has a creſt which reaches to its neck, and which, as well as the whole of the head, being pale yellow, is ſtrongly contraſted with its red muſtachoes ; its appearance is thus remarkable, and the ſoft uncommon colour of its plumage diſtinguiſhes it from the reſt of its genus. The creoles of Cayenne call it the *Yellow Carpenter*. It is ſmaller than the Green Woodpecker, and much more ſlender; its length nine inches. It forms its neſt in large trees, rotten at the core; after it has bored horizontally to the decayed part, it deſcends, and continues the excavation to the depth of a foot and a half. The female lays three eggs, which are white, and almoſt round; and the young are hatched in the beginning of April. The male ſhares the female's ſolicitude, and, during her abſence, he plants himſelf in the entrance. His cry is a whiſtle compoſed of ſix notes, the firſt of which are monotonous, and the two or three laſt flatter. The female has not the bright red bar which appears in the male on each ſide of the head.

There is ſome variety in this ſpecies, certain individuals having all the ſmall coverts of the wings of a fine yellow, and

 the

the great ones edged with that colour; in others, ſuch as that probably which Briſſon deſcribed, the whole plumage is diſcoloured and bleached, ſo as to appear only a dirty white or yellowiſh.

The FERRUGINOUS WOODPECKER, *Lath.*

Le Pic Mordoré. Buff.

Sixth Species.

Picus Cinnamomeus. Gmel.

A FINE bright red, which is brilliant and golden, forms the ſuperb attire of this bird. It is almoſt as large as the Green Woodpecker, but not ſo ſtout. A long yellow creſt in pendulous filaments covers the head, and falls backwards; from the corners of the bill, riſe two muſtachoes of a fine light red, traced nicely between the eye and the throat; ſome white and citron ſpots embelliſh and variegate the rufous ground of the middle of the upper ſurface; the rump is yellow, and the tail black. The female, both of this ſpecies, and of the Yellow Woodpecker which comes from the ſame country, has no red on the cheeks.

The BLACK-BREASTED WOOD-PECKER.

Le Pic a Cravate Noire. Buff.

Seventh Species.

Picus Multicolor. Gmel.

THIS is alſo one of the Yellow *Carpenters* of Cayenne. It has a fine black horſe-ſhoe, which meets the neck behind, covers all the forepart like a cravat, and falls on the breaſt; the reſt of the under-ſide of the body is ruſty fulvous, and alſo the throat and the whole head, whoſe creſt extends to the neck; the back is of a bright rufous; the wing is of the ſame colour, but the quills croſſed with a few black ſtreaks pretty much aſunder, and ſome of theſe extend to the tail, which has a black tip. This Cayenne bird is as large as the Yellow Woodpecker, or even the Ferruginous Woodpecker: all the three are alike ſlender, and ſimilarly creſted. The natives of Guiana give them the common name of *toucoumari*. It appears that the Black-breaſted Woodpeckers lead a life as laborious as the others, and that they inhabit St. Domingo alſo; for Father

 Charlevoix

Charlevoix aſſures us, that the wood employed for building in that iſland is often found bored ſo much by theſe wild carpenters, as to be unfit for uſe *.

The RUFOUS WOODPECKER.

Eighth Species.

Picus Rufus. Gmel.

THE plumage of this little Woodpecker has a ſingular property, viz. the under ſide of its body is of a deeper hue than the upper, contrary to what is obſerved in all other birds. The ground colour is rufous, of various intenſity; deep on the wings; more dilute on the rump and back, more charged on the breaſt and belly, and mingled, on all the body, with black waves, which are very crowded, and which have the effect of the moſt beautiful enamel: the head is rufous, embelliſhed and croſſed by ſmall black waves. This Woodpecker, which is found in Cayenne, is ſcarcely larger than the Wryneck, but it is rather thicker: its plumage, though it conſiſts of

* Hiſtoire de l'îſle de Saint Dominique, par le P. Charlevoix. Paris, 1730, t. 1. p. 29.

only

only two dull tints, is one of the moſt beautiful and moſt agreeably variegated [A].

[A] Specific character of the *Picus Rufus*: "It is rufous, waved with black."

The LITTLE YELLOW-THROATED WOODPECKER.

Ninth Species.

Picus Chlorocephalus. Gmel.

Picus Icterocephalus. Lath.

THIS Woodpecker is not larger than the Wryneck. The ground colour of is plumage is brown tinged with olive, and having white ſpots or ſcales on the fore-part of the body, and under the neck, which is ſpread with a fine yellow that ſtretches under the eye, and on the top of the neck; a red hood covers the crown of the head, and a muſtachoe of that colour diluted riſes from the corners of the bill. This Woodpecker is, as well as the preceding, found in Cayenne [A].

[A] Specific character of the *Picus Chlorocephalus*: "it is olive, below ſpotted with white; its neck and its half creſted head bright yellow; its top red."

The LEAST WOODPECKER of Cayenne.

Tenth Species.

Yunx Minutiſſimus. Gmel.
Picus Cayanenſis Minor. Briſſ.
Picus Minutus. Lath. Ind.
The Minute Woodpecker. Lath. Syn.

THIS bird, as ſmall as the gold creſted wren, is the dwarf of the large family of Woodpeckers. It is not a creeper, and its ſtraight ſquare bill ſhews it to be a real Woodpecker. Its neck and breaſt are waved diſtinctly with black and white zones; its back is brown, ſpotted with white drops, and ſhaded with black; the ſame ſpots, only cloſer and finer marked, appear on the beautiful black that covers the arch of the neck; and laſtly, a little gold head makes it look as handſome, as it is delicate. This little bird, at leaſt if we judge from the ſtuffed ſpecimen, muſt be more ſprightly and agile than any of the other Woodpeckers; and nature would ſeem to have thus compenſated for its ſmallneſs. It is often found in company with the creepers, and like

like them it clambers on the trunks of trees, and hangs by the branches [A].

[A] Gmelin makes this bird a ſecond ſpecies of wryneck under the name of *Yunx Minutiſſimus*: its ſpecific character: "Above, it is black cinerous, below dirty white.".

The GOLD-WINGED WOODPECKER.

Eleventh Species.

Picus Auratus. Linn. Gmel and Borowſk.
Picus Canadenſis Striatus. Briſſ.
Cuculus alis deauratis. Klein.

THOUGH I place this beautiful bird in the cloſe of the family of the Green Woodpecker, I muſt remark, that it ſeems to emerge from even the genus of the Woodpeckers, both by its habits and ſome of its features. Cateſby, who obſerved it in Carolina, tells us, that it is ofteneſt on the ground, and does not creep upon the trunks of trees, but perches on their branches like other birds. Yet its toes are diſpoſed two before, and two behind, like the Woodpeckers; and, like them too, the feathers of its tail are ſtiff and hard; and, what is peculiar to itſelf, the ſide of each is terminated by two ſmall filaments. Its bill

bill is, however, diſſimilar to that of the Woodpeckers; it is not ſquared, but rounded, and ſomewhat curved, pointed, and not formed into an edge. If this bird reſembles then the Woodpeckers in the ſtructure of its feet and tail, it differs in the ſhape of the bill, and in its habitudes, which neceſſarily reſult from the conformation of that principal organ in birds. It ſeems to form an intermediate ſpecies between the Woodpeckers and the Cuckoos, with which ſome naturaliſts have ranged it; and it furniſhes another example of thoſe ſhades by which nature connects her various productions. It is about the ſize of the Green Woodpecker, and is remarkable for its beautiful form, and the elegant diſpoſition of its rich colours; black ſpots, like creſcents and hearts, are ſcattered on the ſtomach and belly on a white ground of a dingy caſt; the forepart of the neck is vinous cinereous or lilack, and, on the middle of the breaſt, there is a broad black zone, ſhaped like a creſcent; the rump is white; the tail black above, and lined below with a fine yellow reſembling dead leaves; the upper ſide of the head, and the top of the neck, are of a lead gray, and the back of the head is marked with a fine ſcarlet ſpot; from the corners of the bill two large black muſta-

choes

No. 163

THE GREAT BLACK WOODPECKER.

choes take their origin, and descend on the sides of the neck, and they are wanting in the female; the back is of a brown ground, with black streaks; the great quills of the wing are of the same cast; but what decorates it, and suffices alone to discriminate the bird, the shafts of all these quills are of a gold colour. It is found in Canada and Virginia, as well as in Carolina [A].

[A] Specific character of the *Picus Auratus*: "It is streaked transversely with gray and black, its throat and breast are black, its nape red, its rump white."

The BLACK WOODPECKER.*

Picus Martius. Linn. and Gmel.

Picus Niger. Briss. Klein and Fris.

Picus Maximus. Ray and Will.

The Great Black Woodpecker. Alb. and Lath.

THIS second species of European Woodpecker appears to be confined to some particular countries, and especially to Germany. However, the Greeks, as well as we,

* In Italian *Picchio, Sgiaia*: In German, *Holtz Krähe*, or Woodcrow, and *Krähe-Specht, Gross-Specht, Schwartz-Specht*, or the crow, the large, the black spight or woodpecker: In Swedish, *Spill-Kraoka*: In Norwegian, *Sort Spæt, Træpikke, Lie Hast*: In Polish, *Dzieciol Naywiekszy*.

we, were acquainted with three ſpecies of Woodpeckers, and Ariſtotle mentions them all †: The firſt, ſays he, is ſmaller than the Blackbird, and is our variegated Woodpecker; the ſecond is larger than the Blackbird, and is the κολιὸς, or our Green Woodpecker ‡; and the third, he repreſents as equal in ſize to a hen, which muſt be underſtood of its length, and not of its thickneſs; and it is therefore our Black Woodpecker, the largeſt of all the Woodpeckers of the antient continent. It is ſixteen inches long, from the tip of the bill to the end of the tail; the bill meaſures two inches and a half, and is of a horn colour; a bright red hood covers the crown of the head; the plumage of the whole body is deep black. The German names *Krähe-ſpecht* and *Holtz-krahe*, crow ſpight and woodcrow, mark both its colour and its ſize.

It is found in the tall foreſts on the mountains of Germany, in Switzerland, and in the Voſges: it is unknown in moſt of the provinces of France ||, and ſeldom deſcends into the low country. Willughby

† *Hiſt. Anim.* Lib. ix. 9.

‡ *Id.* Lib. viii. 3.

|| "The Black Woodpecker is not found in Normandy, nor in the vicinity of Paris, nor in the Orleanois. *Salerne.*

aſſures

aſſures us, that it never occurs in England; and indeed that country is too open for a bird of ſuch a nature, and for the ſame reaſon, it has deſerted Holland* : And this is evidently not on account of the cold of thoſe regions, ſince it inhabits the foreſts of Sweden †. But it is difficult to imagine why it is not found in Italy, as Aldrovandus aſſerts.

Even in the ſame country, theſe birds prefer particular diſtricts that are ſolitary and wild; Friſch mentions a foreſt in Franconia ‡, noted for the multitude of Black Woodpeckers which it contains. In general, the ſpecies is not numerous; and, in the extent of half a league, we can ſeldom find more than a ſingle pair. They ſettle in a certain ſpot, which they ſcarce ever leave.

This bird ſtrikes the trees with ſuch force, that according to Friſch, it may be heard as far as a hatchet. It bores to the heart of the trunk, and forms a very capacious cavity; as much as a buſhel of wood-duſt and chips is often ſeen on the ground below its hole; and ſometimes it hollows out the ſubſtance of the trees to ſuch a degree that they are ſoon borne

* Aldrovandus.

† *Fauna Suecica.* No. 79.

‡ The foreſt of Speſſert.

down

down by the wind ||. They prefer the decayed trees, but, as they alſo attack thoſe which are ſound, the careful proprietors of woods are at pains to deſtroy them. M. Deſlandes, in his Eſſay on the Ship-Building of the Ancients, regrets, that there are few trees fit for making oars forty feet long, which are not bored by the Woodpeckers §.

The Black Woodpecker lays, in the bottom of its hole *, two or three eggs, which are white; as in all birds of the genus, according to Willughby: It ſeldom alights on the ground; the ancients affirmed even, that no Woodpecker ever deſcends from its tree †: When they clamber, the long hind toe is ſometimes placed ſidewiſe, and ſometimes forwards, and is moveable in its joint with the foot, ſo as accommodate itſelf to every poſition: This power is common to all the Woodpeckers.

After the Black Woodpecker has perforated into the cavity of the tree, it gives

|| Ariſtotle, *Hiſt. Anim.* Lib. ix. 9.

§ But M. Deſlandes is much miſtaken in the ſame place, when he ſays that this Woodpecker employs its tongue like an augre to bore the largeſt trees.

* Pliny has aſſerted with too great latitude that the Woodpeckers are the only birds which breed in hollow trees (Lib. x. 18); many other ſmall birds, ſuch as the Titmice, do the ſame.

† Ariſtotle, *Hiſt. Anim.* Lib. ix. 9.

a loud

a loud ſhrill and lengthened ſcream, which is audible at a great diſtance. It alſo makes at times a cracking, or rather a ſcraping, by rubbing its bill rapidly againſt the ſides of its hole.

The female differs from the male in its colour, being of a lighter black, and having no red but on the back of the head, and ſometimes none at all. It is obſerved that the red deſcends lower on the nape of the neck in ſome individuals, and theſe are old males.

The Black Woodpecker diſappears during winter. Agricola ſuppoſed that it remained concealed in hollow trees *: but Friſch affirms, that it retires before the rigour of the ſeaſon, when its proviſions fail; for, continues he, the worms then ſink deep into the wood, and the ant-hills are covered with ice and ſnow.

We know not of any bird of the ancient continent, whether in Aſia or Africa, that is analogous to the European Woodpecker; and it would ſeem to have migrated hither from the New World, where many ſpecies occur that cloſely reſemble it. I proceed to enumerate theſe [A]:—

* *Apud Geſnerum*, p. 677.

[A] Specific character of the *Picus Martius*, Linn. "It is black with a crimſon cap."

BIRDS of the New Continent,

Which are related to the

BLACK WOODPECKER.

The WHITE BILLED WOODPECKER.

Le Grand Pic Noir a Bec Blanc. Buff. *

First Species.

Picus Principalis. Linn. and Gmel.

Picus Carolinensis Cristatus. Briss. †

THIS Woodpecker is found in Carolina, and is the largest of the genus, being equal or even superior in bulk, to the crow. Its bill is white like ivory, and three inches long, channeled through its whole length, and so sharp and strong, says Catesby, that, in an hour or two, the bird often makes a bushel of chips: Hence the Spaniards term it *carpenteros*, or carpenter.

* The Great Black Woodpecker with a white bill.

† Brisson probably measured a very small specimen, when he stated the length of this Woodpecker at sixteen inches; that in the Royal Cabinet, figured in the *Illumined Plates*, was eighteen inches.

Its

Its head is decorated behind by a great ſcarlet tuft, parted into two tufts, one of which falls on the neck, and the other is raiſed, and covered by long black threads, which riſe from the crown and inveſt the whole head, for the ſcarlet feathers lie behind: a white ſtripe, deſcending on the ſide of the neck, and making an angle on the ſhoulder, runs into the white that covers the lower part of the back and the middle quills of the wing; all the reſt of the plumage is a jet deep black.

It hollows its neſt in the largeſt trees, and breeds during the rainy ſeaſon. It is found, too, in hotter climates than that of Carolina; for we recogniſe it in the *picus imbrifœtus* of Nieremberg ‡, and the *quatotomomi* of Fernandez §, though there are ſome differences which would ſeem to indicate a variety ‖; its white bill, three

‡ P. 223.

§ *Hiſt. Nov. Hiſp.* p. 50. *cap.* 186.

‖ The Quatotomomi a kind of Woodpecker of the bulk of a hoopoe; it is variegated with black and fulvous; its bill with which it hollows and bores trees, is three inches long, ſtout, and bright white . . . Its head is decorated by a red creſt three inches long, but its upper part black on either ſide of the neck, a bright white bar deſcends to near the breaſt. It inhabits *Tototepeco*, in higher *Miſteca* not far from the South Sea. It neſtles in lofty trees: feeds on the graſhoppers, *tlaolli*, and on ſmall worms. It breeds in the rainy ſeaſon, that is, between May and September:" Fernandez, *Hiſt. Nov. Hiſp.* p. 50. cap. 186.

inches

inches in length, ſuffices to diſcriminate it. This Woodpecker, ſays Fernandez, inhabits the regions bordering on the South Sea: The North Americans work the bills into coronets for their warriors, and as they cannot procure theſe in their own country, they buy them of the more ſouthern Indians, at the rate of three deer-ſkins for each bill [A].

[A] Specific character of the *Picus Principalis*: It is black, its creſt ſcarlet, a line on either ſide of the neck; the ſecondary feathers of its tail, white." It is a ſcarce bird in North America, and never penetrates beyond the Jerſeys. It breeds in a winding hole, the better to ſcreen the young from the inſinuating rains.

The PILEATED WOODPECKER.

Le Pic Noir a Huppe Rouge. Buff. *

Second Species.

Picus Pileatus. Linn. and Gmel.
Picus Niger Virginianus Criſtatus. Briſſ.
Picus Niger, toto capite rubro. Klein.
The Larger red-creſted Woodpecker. Cateſby.

THIS Wooodpecker, which is common in Louiſiana, occurs equally in Carolina and in Virginia: It reſembles much the preceding, but its bill is not white, and it is

* The Red-creſted black Woodpecker.

is rather ſmaller; though it ſomewhat exceeds the Black Woodpecker of Europe. The crown of the head, as far as the eyes, is decorated by a large ſcarlet creſt, collected into a ſingle tuft, and thrown backwards in the ſhape of flame; above there is a black bar, in which the eye is placed; a red muſtachoe is traced from the root of the bill on the black ſides of the head; the throat is white; a fillet of the ſame colour paſſes between the eye and the muſtachoe, and extends on the neck as far as the ſhoulder; all the reſt of the body is black, with ſome ſlight marks of white on the wing, and a larger ſpot of that colour on the middle of the back; under the body, the black is lighter, and mixed with gray waves; in the female, the forepart of the head is brown, and there are no red feathers, but on the hind part of the head.

Cateſby ſays that theſe birds, not content with rotten trees which ſupply their uſual food, attack alſo the plants of maize; and do much injury, for the wet inſinuates into the holes which they make in the huſk, and ſpoils the ſeeds. But is their motive not to get ſome kind of worms that lurk in the ear, ſince no bird of this genus feeds on grain?

 With

With this bird, we muſt alſo join a Woodpecker which Commerſon brought from the country contiguous to the Straits of Magellan: Its bulk is the ſame, and its other characters pretty ſimilar; only it has no red, except on the cheeks and the fore-part of the head, and the back of its head bears a tuft of black feathers. Thus, the ſame ſpecies occurs, in the correſponding latitudes at the two extremities of the great continent of America. Commerſon remarks that this bird has a very ſtrong voice, and leads a very laborious life; a character that belongs to all the Woodpeckers, which are enured to toil and hardſhip [A].

[A] Specific character of the *Picus Pileatus*: "It is black, its creſt red, its temples and wings marked with white ſpots." It is half the weight of the preceding ſpecies: It ſpreads over the whole extent of North America: lays ſix eggs, and hatches in June. The Indians decorate their calumets with its ſcarlet tuft.

The LINEATED WOODPECKER.

* *L'Ouantou, ou Pic Noir Huppé de Cayenne.* Buff.

Third Species.

Picus Lineatus. Linn. and Gmel.
Picus Varius Brasiliensis. Ray.
Picus Niger Cayanensis Cristatus. Briss.

THIS bird is the *ooantoo* of the Americans, which Barrere has inaccurately pronounced *ventoo*, and the *hipecoo* of Marcgrave. It is as long as the Green Woodpecker; but not so thick; its upper surface is entirely black, except a white line which, rising from the upper mandible, descends like a cincture on the neck, and strews some white feathers on the coverts of the wings; the stomach and belly are waved with black and white bars, and the throat is speckled with the same; from the lower mandible, proceeds a red mustachoe; a beautiful crest of the same colour covers the head and falls backwards; lastly, under the long threads of this crest,

* The Oantoo, or Crested Black Woodpecker of Cayenne.

 we

we perceive ſmall feathers of the ſame red which clothe the top of the neck.

Barrere is right in referring this Woodpecker to the *hipecoo* of Marcgrave, as much as Briſſon is wrong in referring it to the great Carolina Woodpecker of Cateſby: the latter is larger than a crow, and the hipecoo exceeds not a pigeon. And the reſt of Maregrave's deſcription agrees with the ouantoo as much as with the great Carolina Woodpecker, which has not the underſide of its body variegated with black and white, as the ooantoo and the hipecoo; and its bill is three lines, not ſix. But theſe characters belong as little to the Black Woodpecker of Louiſiana; and Briſſon was miſtaken too in placing with it the ooantoo, which as we have juſt ſeen, is nothing but the hipecoo, and would have been better ranged with his eleventh ſpecies.

The ooanto of Cayenne is alſo the *tlauhquechultototl* of New Spain, deſcribed by Fernandez. It bores trees, its head and the upper part of its neck are covered with red feathers. But there is a circumſtance accidentally introduced in his account which ſeems to diſcriminate it from the other Woodpeckers: " The red feathers on the top of the neck, if applied or rather glued to the head, relieve a head-ach; whether

whether this was learnt from experience, or was ſuggeſted by ſeeing *them glued near the head of the bird* [A].

[A] Specific character of the *Picus Lineatus*: "It is black, its creſt crimſon; a white line on either ſide of the neck from the bill to the middle of the back."

The RED NECKED WOODPECKER.

Fourth Species.

Picus Rubricollis Gmel.

THIS bird has not only its head red, but its neck as far as the breaſt of the ſame beautiful colour. It is rather longer than the Green Woodpecker, its neck and tail being elongated, which makes its body appear leſs thick, all the head and neck is covered with feathers to the breaſt, where the tints of that colour melt into the fine fulvous that covers the breaſt, the belly and the ſides; the reſt of the belly is deep brown, almoſt black where the fulvous mixes with the quills of

 the

the wing.—This bird is found in Guiana, as well as the preceding and the following ones [A].

[A] Specific character of the *Picus Rubricollis*: "It is brown, below fulvous; its crested head, and its neck, ferruginous."

The LESSER BLACK WOODPECKER.

Fifth Species.

Picus Flavipes Gmel.
Picus Hirundinaceus. Linn. and Gmel.
Picus Niger Novæ Angliæ. Briss.
Picus Niger Minimus. Klein.
The Yellow legged Woodpecker. Penn.

THIS is the smallest of all the Black Woodpeckers, being only of the size of the Wryneck. A deep black with blue reflections, covers the throat, the breast, the back and the head, except a red spot found on the head of the male; it has also a slight trace of white on the eye, and some small yellow feathers near the back of the head; below the body and along the sternum, there extends a bar of a fine poppy-red; it terminates at the belly, which like

like the ſides is well enameled with black and light gray; the tail is black.

There is a variety of this Woodpecker, which, inſtead of the red ſpot on the crown of the head; has a yellowiſh crown compleatly encircling it; and this is the opening of thoſe ſmall yellow feathers ſeen in the former, and probably reſults from age.—The female has neither a red ſpot nor a yellow circle on the head.

To this ſpecies, we ſhall refer the *leſſer black creeper* of Albin, which Briſſon makes his ſeventh ſpecies, under the name of *the Black Woodpecker of New England* *.

* Mr. Pennant reckons the authority of Albin very ſuſpicious.

The RED-HEADED WOODPECKER.

Le Pic Noir a Domino Rouge. Buff.

Sixth Species.

Picus Erythrocephalus. Linn. Gmel. and Briſſ.
Picus capite colloque rubris. Klein.

THIS bird deſcribed by Cateſby, is found in Virginia: It is nearly as large as the variegated Woodpecker of Europe. Its whole

whole head is enveloped in a beautiful red *domino*, which is ſilky and gloſſy, and falls on the neck; all the under ſurface of the body and the rump are white, and ſo are the ſmall quills of the wing, of which the black joins that of the tail, to form, on the lower part of the back, a great white ſpace; the reſt is black, and alſo the great quills of the wings, and all thoſe of the tail.

Very few of theſe birds are ſeen in Virginia during winter; there are more of them in that ſeaſon in Carolina, though fewer than in ſummer; it would ſeem that they retire to the South to eſcape the cold. Thoſe which remain approach the villages, and even rap on the windows of the houſes. Cateſby adds that this Woodpecker conſumes much fruit and grain; but it probably never recurs to theſe, unleſs in caſe of want of other ſuſtenance, elſe it would differ from all the reſt of the genus.

The GREATER SPOTTED WOODPECKER.

L' Epeiche ou Le Pic Varié. Buff.

First Species.

Picus Major. Linn. Gmel. Bor. Kram. Scopo.
Picus Varius Major. Ray. Will. and Briss.
Picus Discolor. Fris. *

THIS is the third species of the European Woodpeckers. Its plumage is agreeably variegated with white and black, embellished with red on the head and belly. The crown of the head is black, with a red bar on the occiput, and the hood terminates in a black point on the neck; thence rise two branches of black, one of which stretches on each side to the root of the bill and marks a mustachoe, and the other descending to the lower part of the neck, decorates it with a collar; this black streak unites near the shoulder with the black piece that

* In Greek Πιπρα. In Italian *Culrosso*: In German *Bunt-Specht*, *Veiss-Specht*, and *Elsterspech*, from which the French name *Epeiche* is derived. In Swiss *Ægerst-Specht*: In Danish *Flag-Spaet*: In Swedish *Gyllenrenna*: In Norwegian *Kraak-Spinte*: and in Polish *Dzieciol Pstry Wieksly*.

occupies

occupies the middle of the back; two great white ſpaces cover the ſhoulders; in each wing the great quills are brown, the others black, and all mixed with white; the whole of the black is deep, and the whole of the white is pure and unmixed; the red on the head is bright, and that of the belly is a fine ſcarlet. Thus the plumage of this bird is charmingly diverſified, and ſurpaſſes that of all the other Woodpeckers in beauty.

This deſcription anſwers only to the male exactly; the female figured in the *Planches Enluminées* has no red on the back of the head. Some ſpotted Woodpeckers are clothed with a leſs beautiful plumage, and ſome even are entirely white. There is alſo a variety whoſe colours appear more obſcure, and though all the upper ſide of the head and the belly are red, the tint is pale and dull.

Of this variety, Briſſon makes his ſecond *Variegated Woodpecker*, though he had before produced it under the name of the *Great Variegated Woodpecker*; yet theſe two birds are both nearly of the ſame ſize, and have ever been referred to the ſame ſpecies. Belon, who lived in an age when the rules of nomenclature and the errors of ſyſtem had not multiplied the diviſions in the arrangement

rangement of natural objects, classes all these varieties with his *epeiche* or Variegated Woodpecker. But Aldrovandus justly blames both him and Turner for applying to that bird the name *Picus Martius*, which belongs only to the Green Woodpecker.

The Variegated Woodpecker strikes against the trees with brisker and harder blows than the Green Woodpecker; it creeps with great ease upwards, or downwards, and horizontally under the branches; the stiff quills of its tail serve to support it, when it hangs in an inverted posture, and knocks keenly with its bill. It is a shy bird; for when it perceives a person, it hides itself behind a branch and remains still. Like the other Woodpeckers, it breeds in a hollow tree. In our provinces, it approaches the habitations during winter, and seeks to settle on the bark of fruit trees, where the crysalids and eggs of insects are deposited in greater quantity than on the trees of the forest.

In summer during droughts, the Variegated Woodpeckers are often killed at the wood meres, whither these birds repair to drink: it approaches the spot in silence, fluttering from tree to tree; and each time it halts it seems anxiously to examine if any danger

danger threatens: it has an air of inquietude; it liſtens, and turns its head on all ſides, and even looks through the foliage to the ground below; and the leaſt noiſe is ſufficient to drive it back. When it reaches the tree next the mere, it deſcends from branch to branch, until it gets to the loweſt on the margin of the water; it then dips its bill, and at each ſip, it hearkens, and caſts a look round it. After its thirſt is quenched, it retires quickly, without making a pauſe as on its arrival. When it is ſhot on a tree, it ſeldom drops; but as long as a ſpark of life remains, it clings firmly with its nails, ſo that one is often obliged to fire a ſecond time.

This bird has a very large *ſternum*; its inteſtinal canal is ſixteen inches long, but there is no *cæcum*; its ſtomach is membranous; the point of its bill is bony, and five lines in length. An adult male which had been taken from a neſt of five young, weighed two ounces and an half; theſe weighed three gros each, and their toes were diſpoſed as in the father; their bill wanted the two lateral ridges, which, in the adult, took their origin beyond the noſtrils, went below them, and extended two thirds of the length of the bill; the nails, though yet

yet white, were already much hooked. The neſt was in an old hollow aſpin, thirty feet above the ground.

[A] Specific character of the *Picus Major*. "It is variegated with black and white, its vent and the back of its head, red." It is found even in the moſt northern parts of Europe. Its length nine inches, its weight 2 3/4 ounces.

The LESSER SPOTTED WOODPECKER.

Le Petit Epeiche. Buff.

Second Species.

Picus Minor. Linn. Gmel Bor Kram. &c.
Picus Varius Minor. Aldrov. Briſs. and Gerini.
Picus Diſcolor Minor. Friſch and Klein.*

THIS ſpecies reſembles the former ſo cloſely, that it might be regarded as the ſame formed on a ſmaller ſcale, only the fore part of its body is dirty white or rather gray; and it wants the red under the tail, and the white on the ſhoulders. As in the large ſpecies too, it is the male only that has its head marked with red.

* In Italian *Pipra or Pipo*. In German *Speehtle, Graſs Specht*. In Norwegian *Lille, Trœ-pikke* : and in Poliſh *Dzieciol Pſtry Mnieyszy*.

This

This little ſpotted Woodpecker is ſcarcely ſo large as a ſparrow, and weighs only an ounce. In winter it reſorts near houſes and vineyards. It does not creep very high on large trees, and ſeems to prefer the circumference of the trunk.* It neſtles in ſome hole of a tree, and often diſputes the poſſeſſion with the Colemouſe, which is commonly worſted in the ſtruggle and compelled to ſurrender its lodging . It is found in England, where it has received the name of *hickwall*. It alſo inhabits Sweden : and this ſpecies, like that of the greater ſpotted Woodpecker, would appear to be diffuſed even to North-America ; for in Louiſiana a ſmall ſpotted Woodpecker is ſeen which reſembles it almoſt entirely, except that the upper ſide of the head, as in the Variegated Woodpecker of Canada, is covered with a black cap, edged with white.

Salerne ſays that this bird is unknown in France, yet it occurs in moſt of our provinces. The miſtake originated from his confounding the leſſer ſpotted Woodpecker with the Wall Creeper, with which he owns that he was unacquainted. He was equally deceived in aſſerting that Friſch makes no mention of this Woodpecker, from which he

* Geſner.

infers

infers that it exiſts not in Germany: for that naturaliſt ſays only that it is rare, but gives two excellent figures of it.

M. Sonnerat ſaw, in the iſland of Antigua, a ſmall Variegated Woodpecker, which we ſhall refer to this, ſince the characters which he gives are inſufficient to diſcriminate two ſpecies. It is of the ſame bulk; black, ſtriped and ſtreaked with white, covers all the upper ſurface of the body; the under ſurface is ſpotted with blackiſh, on a pale yellow, or rather yellowiſh white ground; a white line marks the ſides of the neck. M. Sonnerat did not perceive red on the head, but he remarks that it was perhaps a female.

[A] Specific character of the *Picus Minor*: It is variegated with white and black, its top red, its vent brick-coloured.

BIRDS

BIRDS of the Ancient Continent,

Which are related to the

SPOTTED WOODPECKER.

The NUBIAN WOOD-PECKER. Lath.

* *L'Epeiche de Nubie Ondé et Tacheté.* Buff.

First Species.

Picus Nubicus. Gmel.

THIS bird is a third ſmaller than the ſpotted Woodpecker of Europe: all its plumage is agreeably variegated with drops and waves broken and, as it were, vermiculated with white and ruſty on a gray brown ground, and blackiſh on the back, and tears of blackiſh on the whitiſh complexion of the breaſt and belly; a half creſt of fine red covers, like a cowl, the back of the head; the crown and the fore-part conſiſt of delicate black feathers, each tip'd with a ſmall white drop; the tail is divided horizontally by brown and ruſty waves. The bird is very handſome, and the ſpecies is new.

* The waved and Spotted Woodpecker of Nubia.

The

The GREAT VARIEGATED WOODPECKER,

Of the Isle of Luçon.

Second Species.

Picus Cardinalis. Gmel.

The Cardinal Woodpecker. Lath.

THIS bird, which is described by Sonnerat, is as large as the Green Woodpecker: the feathers of the back and the coverts of the wing are black, but their shafts are yellow, and there are also yellowish spots on the latter; the small coverts of the wing are striped transversely with white; the breast and the belly are variegated with longitudinal black spots on a white ground; there is a white bar on the side of the neck, extending below the eye; the crown and back of the head, are of a bright red, and for this reason, Sonnerat would apply to it the epithet of *Cardinal*; but the red hood is rather a generie than a specific character, and therefore the name

 which

which that traveller would impoſe, is not ſufficiently deſcriptive. [A]

[A] Specific character of the *Picus-Cardinalis*. "It is black; below, white, ſpotted with black, its top and the back of its head red."

The LITTLE BROWN SPOTTED WOODPECKER,

Of the Moluccas.

Third Species.

Picus Moluccenſis. Gmel.
The Brown Woodpecker. Lath.

THIS little Woodpecker has only two dull and faint ſhades; its plumage is blackiſh-brown, waved with white on the upper ſide of the body, whitiſh ſpotted with brown ſpeckles below; the head and tail, and alſo the quills of the wings, are all brown. It is hardly ſo large as the leſſer ſpotted Woodpecker.

BIRDS

BIRDS of the New Continent,

Which are related to the

SPOTTED WOODPECKER.

The SPOTTED WOODPECKER, Of Canada.

First Species.

Picus Canadensis. Gmel.
Picus Varius Canadensis. Briss.

THIS bird is of the ſame ſize with the European Spotted Woodpecker, and differs only in the diſtribution of its colours. It has no red; and the ſpace which encircles the eye is not white, but black: there is more white on the ſide of the neck, and white or faint yellow on the back of the head. Theſe differences however are ſlight, and the two contiguous ſpecies are perhaps the ſame, only altered by the change of climate.

The *Quauhtotopotli alter* of Fernandez, which is a Woodpecker variegated with black and white, appears to be the ſame with this Canadian Woodpecker; eſpecially as that author never mentions its having any red,

 and

and ſeems to inſinuate that it comes to New Spain from the North. And there muſt be Spotted Woodpeckers in thoſe tracts, ſince travellers have found them in the iſthmus of Darien [A].

[A] Specific character of the *Picus Canadenſis*: "It is white; its top, its back, its ſhoulders, and the two middle quills of its tail, black; the reſt, and the wings, variegated with white and black."

The VARIED WOODPECKER.

L' Epeiche du Mexique. Buff.

Second Species.

Picus Tricolor. Gmel.

I AM much inclined to think that the *great Variegated Mexican Woodpecker* of Briſſon, and his *Little Variegated Mexican Woodpecker*, are the ſame bird. He borrows the firſt from Seba, on whoſe authority Klein and Moehring have inſerted it in their ſyſtems: but it is well known how inaccurate are moſt of the deſcriptions of that compiler. Klein mentions the ſame bird twice, and it is one of thoſe which we have rejected from the family of Woodpeckers. On the other hand, Briſſon, for a reaſon which we cannot gueſs, applies

applies to his ſecond Mexican Woodpecker the epithet *little*, though Fernandez, the only original author, ſays that it is *large*, which he repeats twice in four lines. According to him, it is equal in bulk to the Mexican crow; its plumage is varried with white tranſverſe lines on a black and brown ground; the belly and breaſt are vermillion. This Woodpecker inhabits the cooleſt parts of Mexico, and bores the trees like the reſt of the kind.

The VARIEGATED JAMAICA WOODPECKER.

Third Species.

Picus Carolinus. Linn. and Gmel.
Picus Varius Jamaicenſis. Briſſ.
Picus Varius Medius. Sloane.
Picus Varius Medius Jamaicenſis. Ray.
The Jamaica Woodpecker. Edw.
The Red-bellied Woodpecker. Cateſby.
The Carolina Woodpecker. Penn. and Lath.

THIS Woodpecker is of a middle ſize between the Green Woodpecker and the Spotted Woodpecker of Europe: Cateſby makes it too ſmall, when he compares

 it

it to the Spotted Woodpecker; and Edwards repreſents it too large, in aſſerting it to be equal in bulk to the Green Woodpecker. The ſame author reckons only eight quills in the tail, but probably the two others were wanting in the ſubject which he deſcribes; for all the Woodpeckers have ten quills in the tail. It has a red hood which falls on the arch of the neck; the throat and ſtomach are ruſty gray, which runs by degrees into a dull red on the belly; the back is black, ſtriped tranſverſely with gray waves in feſtoons, which are lighter on the wings, broader and entirely white on the rump.

The figure which Sir Hans Sloane has given of it is very defective, and it is the only one that this naturaliſt and Brown found in the iſland of Jamaica, though there are great many others on the continent of America. The preſent occurs alſo in Carolina, and notwithſtanding ſome differences, it may be recognized in the red bellied Woodpecker of Cateſby. The front is a ruſty white, and in the male, red.

[A] Specific character of the *Picus Carolinus*: "Its cap and its nape are red; its back marked with black ſtripes; its middle tail quills, white, dotted with black."

The STRIPED WOODPECKER, of Louiſiana.

Fourth Species.

Picus Carolinus var. Gmel. and Lath.

IT is rather larger than the Spotted Woodpecker; all the upper ſurface is agreeably ſtriped with white and black, diſpoſed in croſs bands; of the quills of the tail, the two exterior and the two middle ones, are mixed with white and black, the reſt are black; all the under ſurface and the fore part of the body are uniform white; gray, and a little dilute red tinges the lower belly. Of two ſpecimens lodged in the royal cabinet, the one has the upperſide of the head wholly red, with ſome ſtreaks of the ſame colour on the throat and under the eyes; the other has its front gray, and no red but on the back of the head, and is probably the female, this being the uſual difference between the ſexes: in both of them, this red is of a feebler and lighter caſt than in the other Variegated Woodpeckers.

The VARIEGATED WOODPECKER of Encenada.

Fifth Species.

Picus Bicolor. Gmel.

THIS bird is not larger than our Leſſer Spotted Woodpecker and is one of the handſomeſt of the genus; its colours are ſimple, but its plumage is richly mailed, and the white and brown gray, with which it is painted, are ſo finely broken and intermingled, as to produce a charming effect. The male has a full creſt, and ſome red feathers appear in it; the female wants the creſt, and its head is entirely brown.

The HAIRY WOODPECKER.

Sixth Species.

Picus Villoſus. Linn. Gmel. and Klein.
Picus Varius Virginianus. Briſſ.

I SHALL borrow the name of *Hairy Woodpecker* from the Engliſh ſettlers of Virginia, becauſe it expreſſes a diſcriminating character of the bird, viz. a white bar, conſiſting

consisting of loose feathers, that extends quite along the back to the rump: the rest of the back is black; the wings too are black, but marked pretty regularly with spots of dull white, round, and in the form of tears; a black spot covers the crown, and red the back of the head, from which a white line extends to the eye, and another is traced on the side of the neck; the tail is black, all the under surface of the body is white. This Woodpecker is rather smaller than the Spotted Woodpecker.

The LITTLE VARIEGATED WOODPECKER, of Virginia. *Buff.*

Seventh Species.

Picus Pubescens. Linn and Gmel.
Picus Varius Virginianus Minor. Briss. and Klein.
The Smallest Woodpecker. Catesby.
The Downy Woodpecker. Lawson and Penn.
The Little Woodpecker. Lath.

We owe to Catesby the account also of this small Woodpecker: it weighs rather more than an ounce and half, and resembles the Hairy Woodpecker so much, it is said, in its spots and colours, that, but

but for the difference of ſize, they might be regarded as of the ſame ſpecies: Its breaſt and belly are light gray, the four middle quills of its tail are black, and the reſt barred with black and white. The female is diſtinguiſhed from the male, as in all the Woodpeckers, by having no red on the head.

The VARIEGATED WOODPECKER, of Carolina.

Eighth Species.

Picus Varius. Linn. and Gmel.
Picus Carolinenſis. Briſs. and Klein.
The yellow-bellied Woodpecker. Cateſby, Penn, and Lath.

THOUGH this Woodpecker has a yellow tinge on the belly, we ſhall not exclude it from thoſe which are variegated with white and black, ſince theſe colours appear on the upper ſurface, which really characterizes the plumage. It is ſcarcely ſo large as the leſſer Woodpecker; all the upper ſide of its head is red; four ſtripes, alternately black and white, cover the ſpace between ſpace between the temple and the cheek, and the laſt of theſe ſtripes bounds the throat, which is of the ſame red with the head; the black and white intermingle and interſect each other agreeably on the back, the wings, and the tail; the fore part of the body is a light yellow, ſprinkled with ſome black ſpeckles. The female wants the red. This Woodpecker inhabits, according to Briſſon, Virginia, Carolina, and Cayenne.

The VARIEGATED UNDATED WOODPECKER,

Ninth Species.

Picus Tridactylus var. Linn. Gmel. &c.
The Southern three-toed Woodpecker, Lath.

THE plumage reſembles that of the Spotted Woodpecker; the back is black, with white diſpoſed in waves or rather ſcales on the great quills of the wing, and theſe two colours form, when it is cloſed, a checked bar: the under ſurface of the body is white, variegated on the ſides with black ſcales; two white ſtreaks ſtretch backwards, one from the eye, the other from the bill, and the top of the head is red.

The

The figure of this bird agrees perfectly with Brisson's description of the *Variegated Cayenne Woodpecker*, except that the former has four toes as usual, and the latter only three. We cannot therefore doubt the existence of three-toed Woodpeckers: Linnæus describes one found in Dalecarlia; Schmidt, one in Siberia; and we are informed by Lottinger* that it occurs also in Switzerland. The three-toed Woodpecker appears therefore to inhabit the north of both continents. Ought the want of the toe to be regarded as a specific character, or considered as only an accidental defect? It would require a great many observations to answer that question; but it may be denied that the same bird inhabits also the equatorial regions, though after Brisson it is termed *the Spotted Cayenne Woodpecker* in *the Planches Enluminees*.

After this long enumeration of the birds of both continents that are akin to the Woodpeckers, we must observe that we have been obliged to reject some species noticed by our nomenclators. These are the third, eighth, and twentieth, which Brisson ranges with the Woodpeckers, Seba with the herons, and Moehring with the crows.

* Extract of a letter from M. Lottinger, to M. de Montbeillard, dated Strasburg, 22d September, 1774.

crows. Klein calls theſe birds *harpooneers*; becauſe, according to Seba, they dart from the air upon fiſh and transfix them. But this habitude belongs not to the Woodpeckers; and the diſpoſition of the toes, which in Seba's figure are diſpoſed *three* and *one*, beſides demonſtrate that they are quite a diſtinct kind.

The WOODPECKER-CREEPERS.

Les Pic-Grimpereaux. Buff.

THE genus of theſe birds, of which we know only two ſpecies, appears manifeſtly diſcriminated, and conſtitutes the intermediate link between the Woodpeckers and the Creepers. The firſt and largeſt ſpecies reſembles moſt the creepers, by its curved bill; and the ſecond, on the contrary, is more analogous to the Woodpeckers ſince it has a ſtraight bill. Both of them have three toes before and one behind, like the Creepers; and at the ſame time the quills of their tail are ſtiff and pointed, like the Woodpeckers.

The firſt was ten inches long; its head and throat ſpotted with rufous and white; the upper ſide of the body, rufous, and the underſide,

underſide, yellow, ſtriped tranſverſely with blackiſh; the bill and feet black.

The ſecond was only ſeven inches long; its head, neck, and breaſt, ſpotted with rufous and white; the underſide of the body is rufous, and the belly ruſty brown; the bill gray, and the feet blackiſh.

Both theſe birds have very nearly the ſame natural habits; they creep againſt trees like Woodpeckers, ſupporting themſelves by the tail; they bore the bark and the wood with much noiſe, and they feed upon the inſects thus detected: they inhabit the foreſts and ſeek the vicinity of ſprings, and rivulets. The two ſpecies live together, and often on the ſame tree, on which many other ſmall birds are perched; yet they are only fond of each others ſociety, and never intermix the breed. They are very agile, and flutter from tree to tree, but never perch or fly to a diſtance. They are commonly found in the interior parts of Guiana, where the natives of the country confound them with the Woodpeckers; which is the reaſon that they have received no appropriated name. It is probable that they alſo inhabit the other warm climates of America, though no traveller has mentioned them.

The

No. 164

THE COMMON WRYNECK.

The WRYNECK.

Le Torcol. Buff.
Yunx-Torquilla. Linn. and Gmel.
Iynx, ſeu Torquilla. Aldrov. Geſner. Will. &c.
Torquilla. Briff *.

THIS bird may be diſtinguiſhed at firſt ſight by a habit peculiar to itſelf; it twiſts and turns back its neck, its head reverted on its back, and its eyes half-ſhut,† and the motion is ſlow, tortuous, and exactly ſimilar to the waving wreaths of a reptile. It ſeems to be occaſioned by a convulſion of ſurprize and fright; and it is alſo an effort which the bird makes to diſengage itſelf when held, yet this motion is natural to it, and depends in a great meaſure on its ſtructure; for the callow brood

* In Greek Ιυγξ; in modern Latin *Torquilla*; and almoſt all its other names in various languages refer to the diſtortion of its neck; in French *Torcol*; in Italian *Tortocollo, Capotorto, Verticella*; in Spaniſh *Torzicuelle*; in German *Wind-halz, Nater-halz, Naterz wang, Nater-wendel*; in Swediſh *Gioek-tita*; in Daniſh, *Bende-halz*; in Norwegian *Sao-gouk*; in Poliſh *Kretoglow*; in Ruſſian *Krutiholowa*.

† Ariſtotle, *Hiſt. Anim.* Lib. ii. xii. Schwenckfeld.

have

have the ſame vermicular wreathing, and many a timorous neſt-finder has fancied them to be young ſerpents ||.

The Wryneck has alſo another ſingular habit. One which had been ſhut twenty-four hours in a cage, turned towards a perſon who approached it, and, eyeing him ſteadily, it roſe upon its ſpurs, ſtretched ſlowly forwards, raiſing the feathers on the top of its head and ſpreading its tail; then it ſuddenly drew back, ſtriking the bottom of the cage with its bill, and retracting its creſt. It repeated this geſture, which was alſo obſerved by Schwencfeld, to the number of an hundred times, and as long as the ſpectator remained beſide it.

Theſe ſtrange attitudes and natural contortions ſeem antiently to have prompted ſuperſtition to adopt this bird in enchantments, and to preſcribe its fleſh as the moſt powerful incentive to love; inſomuch, that the name *Iynx* denoted all ſort of enchantments, violent paſſions, and whatever we call the charm of beauty; that blind power which irreſiſtibly commands our affections. Such is the ſenſe in which Heliodorus, Lycophron, Pindar, Æſchylus, and Sopho-

|| Belon.

cles

cles employ it. The enchantreſs in Theocritus makes this charm, to recall her lover. It was Venus herſelf that, from Mount Olympus, brought the Jynx to Jaſon, and taught him its virtue, to win the heart of Medea, (Pindar). The bird was once a nymph, the daughter of Echo: by her enchantments, Jupiter became enamoured of Aurora; and Juno, in wrath, performed the metamorphoſis. *See* Suidas, and the Scholiaſt of Lycophron, Æſchylus, Sophocles, Heliodorus, Pindar, and Eraſmus.

The ſpecies of the Wryneck is no where numerous: each individual leads a ſequeſtered life, and even migrates ſolitarily. They arrive ſingly in the month of May†. They never enter into any ſociety but that of the female, and it is only tranſitory, for the domeſtic union is diſſolved, and they retire in September. The Wryneck prefers, for the ſake of ſolitude, a ſtraggling tree in the midſt of ſome broad hedge-row. Towards the end of ſummer, it is found alone among the fields of corn, particularly oats, and in the ſmall paths that run through patches of buck-wheat. It feeds on the ground,

† Geſner ſays that he has ſeen them in the month of April.

and does not clamber on the trees like the Woodpeckers, though it is closely related to these birds, and has the same conformation in its bill and feet. Yet it never intermixes with them, and seems to form a small separate family.

The Wryneck is as large as the Lark, being seven inches long, and ten inches across the wings; all its plumage is a mixture of gray, of black, and of tawny, disposed in waves and bars, contrasted so as to produce the richest enamel with these dusky shades ||, the underside of the body is of a white gray ground, tinged with rusty on the neck, and painted with small black zones, which separate on the breast, and stretch into a lance-shape, and are scattered and diluted on the stomach; the tail consists of ten flexible quills, which the bird spreads when it flies, and which are variegated below with black points on a dusky-gray ground, and intersected by two or three broad waved bars like those on the wings of night flies; the same mixture of beautiful undulations of black, of brown, and of gray, among which are perceived zones, lozenges, and zigzag lines, paints all the upper surface on a deeper and more rusty

|| Pindar calls the Jynx, variegated. *Gesner.*

ground.

ground. Some describers have compared the plumage of the Wryneck to that of the Woodcock; but it is more agreeably varied, its tints are clearer and distincter, of a softer feel, and have a finer effect; the cast of the colour is rufous in the male, and more cinereous in the female, which discriminates them; the feet are rusty gray; the nails sharp, the two exterior much longer than the two interior.

This bird holds itself very erect on the branch where it sits, and its body is even bent backwards. It clings in the same way to the trunk of a tree when it sleeps; but it never clambers like the Woodpecker, nor seeks its food under the bark; its bill, which is nine lines in length and fashioned as those of the Woodpeckers, does not assist it in finding its nourishment, and is nothing but the sheath of a large tongue three or four fingers in length, which it darts into the ant-hills, and which it again draws back covered with ants that stick to its viscous humidity. The point of the tongue is sharp and horny; and, to give it extension, two great muscles rise from its root, and, after inclosing the larynx, and stretching to the crown of the head as in the Woodpeckers, are inserted in the front.

It

It alſo wants commonly the *cæcum*†; and Willughby ſays, that in its ſtead, there is only a ſort of inflation of the inteſtines.

The cry of the Wryneck is a very ſhrill drawling whiſtle, which the ancients properly termed *ſtridor*, and the Greek name Ιυγξ ſeems to imitate the ſound‖. It is heard eight or ten days before the cuckoo. It lays in holes of trees, without making any neſt, and on the duſt of rotten wood, which it throws to the bottom of the cavity, by ſtriking the ſides with its bill. It has commonly eight or ten eggs, which are white as ivory‡. The male carries ants to his mate, during incubation; and the young brood againſt the month of June, writhe their neck, and whiſtle loud when one approaches them. They ſoon abandon their lodgment, and acquire no attachment for each other, ſince they ſeparate and diſperſe as ſoon as they can uſe their wings.

† Albin.

‖ Scaliger derives Ιυγξ from Ιυζειν, which occurs in the 17th book of the Iliad, and ſignifies to ſcream.

‡ On the 12th of June, we received ten eggs of a Wryneck, taken out of a hole in an old apple-tree, five feet from the ground, and which reſted on rotten wood; and three years before, Wryneck's eggs were brought to us from the ſame hole.

They can hardly be raiſed in a cage, it being very difficult to procure the proper food. Thoſe which had been kept ſome time, touched the paſte that was offered them with the tip of their tongue, and after taſting, rejected it, and died of hunger § An adult Wryneck, which Geſner tried to feed with ants, lived only five days: it conſtantly refuſed to eat other inſects, and periſhed ſeemingly from languor.

About the end of ſummer, this bird grows very fat, and is then excellent meat; ſo that in many countries it goes by the name of *ortolan*. It is ſometimes caught by the ſpring, and the ſportſmen tear out its tongue with the view to prevent its fleſh from contracting the taſte of ants. The ſeaſon is from Auguſt to the middle of September, when the Wrynecks depart, none of them remaining during the winter in our climates.

§ On the 6oth of June, I cauſed a Wryneck's neſt to be taken out of an old crab five feet from the ground; the male remained on the high branches of the tree, and cried very loud, while his female and his young were diſlodged. I fed them with paſte made of bread and cheeſe; they lived near three weeks: they were familiar with the perſon who took care of them, and would come to eat out of his hand. When they grew large, they refuſed the uſual paſte, and as no inſects could be procured for them, they died of hunger."
Note, communicated by M. Gueneau de Montbeillard.

The ſpecies is ſpread through all Europe, from the ſouthern ſtates to Sweden*, and even Lapland†: it is common in Greece‖, and Italy.= We learn from a paſſage of Philoſtratus‡ that the Wryneck was known to the Magi, and found in the region of Babylon. Edwards aſſures us, that it occurs in Bengal. In ſhort, though its numbers are in each country rare, they are diffuſed through the whole extent of the ancient continent. Aldrovandus alone ſpeaks of a variety in the ſpecies; but his deſcription was made from a drawing, and the differences are ſo ſlight, that we have thought it unneceſſary to ſeparate it. [A]

* Fauna Suecica.

† Rudbeck.

‖ Belon.

= Aldrovandus.

‡ Vita Apollon.

[A] Specific character of the Wryneck, *Yunx-Torquilla.*

"It is variegated with white, gray, and ferruginous." Its Swediſh name *Gioëk-Tita,* ſignifies *the Cuckoo's explainer*: and the Welſh *Gwâs y gog,* means *the Cuckoo's attendant.* In fact the Wryneck uſually appears a little before the Cuckoo. It weighs an ounce and a quarter. Its egg is white, and ſemi-tranſparent.

The

The BARBETS.

NATURALISTS have applied the epithet *bearded* to ſeveral birds that have the baſe of their bill beſet with detached feathers, long and ſtiff like briſtles, all of them directed forward. But we muſt obſerve, that under this deſignation, ſome birds of different ſpecies and from very diſtant climates have been confounded. The *tamatia* of Marcgrave, which comes from Brazil, has been ranged with the Barbet of Africa and that of the Philippines; and all thoſe which have a beard on their bill, and two toes before and two behind, have been grouped together by our nomenclators. Yet the natives of the old continent are diſcriminated from thoſe of the new, their bill being much thicker, ſhorter, and more convex below. We ſhall therefore give the name *tamatia* to the former kinds, and appropriate *barbet* ſolely to the latter.

The

The TAMATIA.

First Species.

Bucco-Tamatia. Gmel.
Tamatia Brasiliensis, Marcg.
Tamatia Guacu. Pison.
The Spotted-bellied Barbet. Lath.

WE have already remarked that Brisson was mistaken in ranging this bird with the little thrushs of Catesby; for it is entirely different in the disposition of its toes, and the shape of its bill, as well as in its beard, and the largeness of its head, a property common to the whole genus. Marcgrave is inaccurate too, in asserting that it wants the tail; this is indeed short, but his specimen was probably incom lete. All the other characters correspond; and as the bird occurs not only in Brazil, but also in Cayenne, whence it has been sent to us, we can easily compare the description.

Its total length is six inches and a half of which the tail occupies two inches; the bill is fifteen lines, its upper extremity hooked, and as it were, divided into two points; the beard which covers it, extends more than half its length; the upper side of the head

No. 165

THE SPOTTED-BELLIED BARBET.

head and the front, are rufty; on the neck, there is an half-collar variegated with black and rufous; all the reft of the plumage above is brown, fhaded with rufous; on each fide of the head behind the eyes, there is a pretty large fpot; the throat is orange, and the reft of the lower furface of the body is fpotted with black on a rufty white ground; the bill and feet are black.

The habitudes of this Tamatia are the fame with thofe of all the birds of the fame genus found in the New Continent: they refide in the moft fequeftered parts of the forefts, and conftantly remote from dwellings and even cleared grounds: they never appear in flocks, or in pairs: they fly laborioufly, and to fhort diftances, and never alight but on low branches, preferring fuch as are thickeft clothed with fprays and leaves: they have little vivacity, and when once feated, they remain a long time: they have even a-dull, melancholy afpect, and they might be faid to affect giving themfelves an air of gravity, by finking their large head between their fhoulders; and it then feems to cover all the forepart of the body. Their difpofition correfponds exactly to their maffy figure and their ferious deportment. They are fo unwieldly that they have much difficulty to move; and

and a perſon may advance as near as he pleaſes and fire ſeveral times, without driving them to flight. Their fleſh is not bad, though they live on caterpillars and other large inſects. In ſhort, they are exceedingly ſilent and ſolitary, homely and remarkably ill ſhaped [A].

[A] Specific character of the *Bucco-Tamatia*: "It is rufous brown; below rufous white, ſpotted with black; its throat orange; it is half-collared."

The TAMATIA with the Head and Throat Red.

Second Species.

Bucco-Cayanenſis. Gmel. and Briſſ.
The Cayenne Barbet. Lath.

THIS bird, which we have delineated in the ſame plate under two different denominations, appears however not to form two ſpecies, but only to include a variety; for in both, the head and throat are red; the cheeks and all the under ſurface of the body black; the bill blackiſh, and the feet cinereous. The only difference is, that in the firſt figure the breaſt is yellowiſh white, but in the ſecond it is brown diluted

diluted with yellow; the former has black ſpots on the top of the breaſt; it has alſo a ſmall white ſpot above the eyes and white ſpots on the wings, which are wanting in the ſecond. But as, in other reſpects, they are ſimilar, and are of the ſame ſize, we do not regard the differences of colour ſufficient to conſtitute two diſtinct ſpecies, as our nomenclators have preſumed. Theſe birds are found not only in Guiana, but in St. Domingo, and probably in other hot climates of America [A].

[A] Specific character of the *Bucco Cayanenſis*: "It is black, below ochry-white, its front and throat red."

The COLLARED TAMATIA

Third Species.

Bucco Capenſis. Linn. and Gmel.
Bucco. Briſſ.
The Collared Barbet. Lath.

THE plumage is agreeably variegated; the underſide of the body deep orange, ſtriped tranſverſely with black lines; about the neck there is a collar, which is very narrow above and ſo broad below as to cover all the top of the breaſt; and this black

black collar is accompanied with a half collar of a tawny colour; the throat is whitish; the lower part of the breast is rusty white, which constantly inclines more to rufous as it descends under the belly; the tail is two inches and three lines in length, and the whole bird measures seven inches and a quarter, its bill an inch and five lines, and the legs which are gray, seven and a half lines in height. It is found in Guiana, but is rare [A].

[A] Specific character of the *Bucco Capensis*: "It is rufous with a fulvous stripe on its shoulder, a black stripe on its breast."

The BEAUTIFUL TAMATIA.

Fourth Species.

Bucco Elegans. Gmel.
Bucco Maynanensis. Briss.
The Beautiful Barbet. Lath.

THIS bird is the most beautiful, or rather the least ugly of the genus: it is better made, smaller and more slender than the rest, and its plumage is so variegated that it would be difficult to give a full description. It is five inches eight lines in length, including the tail which is near two inches;

inches; the bill meaſures ten lines and the legs the ſame.

It is found on the banks of the river Amazons, in the country of the Maynas: but we are informed that it inhabits equally the other parts of South America.

The BLACK and WHITE TAMATIAS.

Fifth and Sixth Species.

WE can ſcarce ſeparate theſe two birds, for they differ only in ſize; and beſides their general reſemblance in the colours, they have both an appropriated character; viz. their bill is ſtronger, thicker and longer than the other Tamatias in proportion to their body; in both alſo the upper mandible is very hooked, and terminates in two points as in the firſt ſpecies.

The largeſt * of theſe Black and White Tamatias is very thick compared with its length, which is hardly ſeven inches: it is a new ſpecies ſent from Cayenne by M. Duval, and alſo the other ſpecies which is

* *Bucco Macrorynchos.* Gmel.
The Greater Pied Barbet. Lath.

ſmaller

ſmaller and exceeds not five inches in length. It would be tedious to enter into a minute deſcription. Their ſimilarity is ſo ſtriking that but for their difference of ſize, we might regard them as the ſame ſpecies. [A]

[A] Specific character: It is black; its front, the tips of its tail-quills, and the under-ſide of its body, are white; it has a black ſtripe on its breaſt."

The BARBETS.

Les Barbus. Buff.

WE applied the name *Tamatia* to the bearded birds of America, and reserved *Barbet* for those of the ancient continent. As both these fly with great difficulty, on account of the thickness and unwieldiness of their body, it is improbable that they could migrate from one continent into another, since they inhabit the hottest climates. Accordingly, they are different, and we have therefore discriminated them. But still they resemble in many characters; for, besides the long slender bristles that cover the bill either wholly or partly and form the beard, and the position of the feet, which is the same in both; they have equally a squat body, a very large head, and a bill exceedingly thick, somewhat curved below, convex above, and compressed on the sides. In the Barbets however, the bill is sensibly shorter, thicker, and rather less convex below than in the Tamatias. Their dispositions also differ; the former are sedate and almost stupid, while the latter, which inhabit the East Indies, attack

attack the ſmall birds, and, in their œconomy, reſemble nearly the Shrikes.

The YELLOW THROATED BARBET.

Firſt Species.

Bucco Philippenſis. Gmel. and Briſſ.

ITS length is ſeven inches; the tail only eighteen lines; the bill twelve or thirteen lines, and the legs eight lines: the head is red, and alſo the breaſt; the eyes are encircled by a large yellow ſpot; the throat is pure yellow, the reſt of the underſide of the body yellowiſh, variegated with longitudinal ſpots of dull green. The female is ſmaller than the male; and has no red on the head or breaſt. They are found in the Philippines [A].

[A] Specific character of the *Bucco Philippenſis*: "It is green, the top of the head (of the male); the ſpace about the eyes, the throat, and the underſide of the body, yellow."

No. 166

THE BARBET.

The BLACK THROATED BARBET.

Second Species.

Bucco Niger. Gmel.

THIS ſpecies, which, as well as the preceding, is found in the Philipinnes, is yet very different. It is thus deſcribed by Sonnerat.

" This bird is rather larger and particularly longer than the Groſbeak of Europe; the front or fore part of the head is of a beautiful red; the crown, the back of the head, the throat, and the neck, are black; there is alſo, above the eye, a ſemi-circular black ſtripe which is continued by another one, ſtraight and white, that deſcends to the lower part of the neck on the ſide; below the yellow ſtripe and the white one, which continues it, there is a black vertical ſtripe, and between this and the throat, there is a white longitudinal ſtripe, that is loſt at its baſe in the breaſt, which, as well as the belly, the ſides, and the upper ſurface of the tail, is white; the middle of the back is black, but the feathers on the ſide, between the neck and the back, are black,

 ſpeckled

ſpeckled each with a ſtreak of yellow; the four firſt, including the ſtump, are tipt with white, and the fifth with yellow, which forms a croſs ſtripe on the top of the wing; below this ſtripe, are black feathers, ſpeckled each a yellow point: the laſt feathers finally, which cover the great quills of the wing, are likewiſe entirely black; but the others have over their whole length, on the ſide where the webs are longeſt, a yellow fringe; the tail is black in the middle, tinged with yellow on the ſides; the bill and feet are blackiſh." [A]

[A] Specific character of the *Bucco Niger*: "It is black, below white, a yellow ſtreak above the eye produced on either ſide to the neck, a white collar."

The BLACK-BREASTED BARBET.

Le Barbu à Plaſtron Noir. Buff.

Third Species.

Bucco Niger. Var. Gmel.

THIS is a new ſpecies ſent to us from the Cape of Good Hope, but without any account of the natural habits of the bird. It is ſix inches and a half long; its tail eighteen

eighteen lines; its feet eight or nine lines. It is of a middle ſize, ſmaller than the Groſbeak of Europe; its plumage is agreeably mingled and contraſted with black and white; its front is red; there is a yellow line on the eye, and drops of a bright ſhining yellow ſcattered on the wings and the back; the ſame yellow tinge extends in daſhes to the rump; and the quills of the tail, and the middle ones of the bill, are ſlightly fringed with the ſame colour; a black plate covers the breaſt as far as the neck; the back of the head is alſo enveloped in black, and a black bar between two white ones deſcends on the ſide of the neck.

The LITTLE BARBET.

Fourth Species.

Bucco Parvus. Gmel.

THIS is a new ſpecies, and the ſmalleſt in the genus. It was given to us as brought from Senegal, but without mention of any other circumſtance. It is only four inches long: its large head and thick bill ſhaded with long briſtles, ſufficiently characterize it. Its tail is ſhort, and its wings, when cloſed, reach almoſt to its

 extremity;

extremity: all the upperſide of the body is of a blackiſh brown, ſhaded with fulvous, and tinged with green on the quills of the wing and of the tail; ſome ſmall white waves form fringes on the former: the underſide of the body is whitiſh, with ſome traces of brown; the throat is yellow, and from the corners of the bill a ſmall white bar paſſes under the eyes. [A]

[A] Specific character of the *Bucco Parvus*: "It is blackiſh brown, below, white, ſpotted with brown; its throat yellow."

The GREAT BARBET.

Fifth Species.

Bucco Grandis. Gmel.

THIS bird is near eleven inches long. The principal colour in its plumage is a fine green, which mingles with other colours on different parts of the body, and eſpecially on the head and neck: the head entirely, and the forepart of the neck, are green mixed with blue; ſo that they appear green or blue, according to their expoſure to the light; the origin of the neck, and that of the back, are of a cheſnut brown, which varies alſo in differ-

ent

ent aſpects, being intermingled with green; all the upperſide of the body is of a very fine green, except the great quills of the wings, which are partly black; all the underſide of the body is of a much lighter green; there are ſome feathers below the tail, which are of a very fine red: the bill is ten lines in length, and an inch broad at the baſe, where there are black hard briſtles; it is whitiſh, but black at the point: the wings are ſhort, and ſcarcely reach the middle of the tail. This bird was ſent to us from China. [A]

[A] Specific character of the *Bucco Grandis*: "It is green, the lower coverts of its tail red."

The GREEN BARBET.

Sixth Species.

Bucco Viridis. Gmel.

It is ſix inches and a half long; the back, the coverts of the wings, and of the tail are of a very fine green; the great quills of the wing are brown, but that colour is not ſeen, being hid by the coverts of the wings; the head is brown gray; the neck is the ſame colour, but each feather is edged with yellowiſh, and above and behind the eye, there is a white ſpot; the belly is of a much paler green than the back; the bill is whitiſh, and the baſe of the upper mandible is ſurrounded with long black and hard hairs; the bill is an inch and two lines in length, and ſeven lines broad at its baſe, the wings are ſhort, and reach only the middle of the tail. We received it from the Eaſt Indies. [A]

[A] Specific character of the *Bucco Viridis*: "It is green, its head and neck gray-brown, the ſpace about its eyes white."

The TOUCANS.

WHAT may be termed the physiognomy of animated beings results from the aspect of their head in different positions. Their form, their figure, their shape, &c. refer to the appearance of their body and of its members. In birds, it is easy to perceive, that such as have a small head and a short slender bill, exhibit a delicate, pleasing, and sprightly physiognomy. Those, on the contrary, with an over-proportioned head, such as the Barbets, or with a bill as large as the head, such as the Toucans, have an air of stupidity, which seldom belies their natural talents. A person who saw a Toucan for the first time might take the head and bill, in the front view, as one of those long-nosed masks that frighten children; but when he seriously examined this enormous production, he would be surprized that nature had given so huge a bill to a bird of such moderate size, and his astonishment would increase on reflecting, that it was useless and even burthensome to its owner, which it obliged to swallow its food whole without dividing or

 crushing.

crushing. And so far is this bill from serving the bird as an instrument of defence, or even as a counterpoise, that it acts like a weight on a lever which tends constantly to destroy the balance, and occasion a sort of hobbling motion.

The true character of nature's errors is disproportion joined to inutility. All those parts of animals which are overgrown or misplaced, and which are useless or inconvenient, ought not to be ranged in the great plan of the universe; they should be imputed to her caprice or oversight, which however tend equally to their object, since these extraordinary productions evince, that whatever can be, actually exists; and that though a regular symmetry commonly obtains, the disorders, the excesses, and the defects which are permitted, demonstrate the extent of that power which is confined not to those ideas of proportion and system that we are apt to regard as the standards.

And as nature has bestowed on most animals all the qualities that conspire to beauty and perfection of form, she has also admitted more than one defect in her careless productions. The enormous and useless bill of the Toucan includes a tongue, still more useless, and whose structure is very

very uncommon; it is not fleſhy or cartilaginous as in other birds, but a real feather miſplaced.

The word *Toucan* means *feather* in the language of Brazil, and *Toucan tabouracé*, which ſignifies *feathers for dancing*, is applied by the natives to that bird which furniſhes them with the decorations worn on feſtival days. Indeed theſe birds, ſo ſhapeleſs and monſtrous in their bill and tongue, have a brilliant plumage; their throat is orange of the moſt vivid hue; and though ſuch beautiful feathers are found only in ſome ſpecies of Toucans, yet have they given name to the whole genus. Theſe feathers are even in Europe in requeſt for making muffs. The huge bill of the Toucan has acquired it other honours, and tranſlated it among their ſouthern conſtellations, where nothing is admitted but what is ſtriking or wonderful * Beſides the exceſſive length of this bill, it is through its whole length wider than the head of the bird; Lery has termed it, *the bill of bills* †, other voyagers have named the Toucan the *bird all-bill* ‡, and the Creoles of Cayenne

* Journal des Obſervations Phyſiques du P. Feuillée, p. 428.

† Voyage du Braſil. p. 174.

‡ Dampier's Voyage round the World.

apply

apply to it the epithet of *Grosbeak.* The magnitude of the bill would exceedingly fatigue the head and neck, were it not very thin, insomuch that it may easily be crushed between the fingers. Authors || were therefore mistaken in asserting, that the Toucans bored trees, like the Woodpeckers; or they were at least mislead by the Spaniards, who have confounded these birds under the same name *carpenteros* (carpenter), or *tacatacas* in the language of Peru. But it is certain that the Toucans are very different from the Woodpeckers, and could not imitate them in that habit; and indeed Scaliger has before us remarked that as their bill was hooked downwards, it seemed impossible for them to make a perforation.

The form of this huge unwieldly bill is very different in each mandible: the upper one is bent into the shape of a sickle, rounded above, and hooked at the extremity; the under one is shorter, narrower and less curved below. Both of them have indentings on the edges, but which are much more perceptible on the upper than on the under; and what appears still more singular, these indentings do not fit into each other, nor even correspond in their relative position, those on the right side not

|| Hernandez.

being

being oppoſite to thoſe on the left. for they begin more or leſs behind, and end alſo more or leſs forward.

The tongue of the Toucans is, if poſſible, more wonderful than the bill: they are the only birds which may be ſaid to have a feather inſtead of a tongue, and a feather it certainly is, though the ſhaft is a cartilaginous ſubſtance two lines broad; for on both ſides there are very cloſe barbs, entirely like thoſe of ordinary feathers, and which are longer the nearer they are inſerted to the extremity. With an organ ſo ſingular and ſo different from the ordinary ſubſtance and organization of the tongue, we might ſuppoſe that theſe birds were mute; yet they have a voice as well as the reſt, and often utter a ſort of whiſtling, which is reiterated ſo quickly, and with ſuch continuance that they have been denominated the *preaching birds*. The ſavages aſcribe great virtues to this feathery tongue *, and uſe it as a cure in many diſorders. Some authors † have ſuppoſed that the Toucans had no noſtrils;

* M. de la Condamine ſpeaks of a Toucan which he ſaw on the banks of the Maragnon, whoſe monſtrous bill is red and yellow; its tongue, he ſays, which reſembles a delicate feather, is eſteemed to have great virtues. *Voyage a la riviere des Amazones, Paris*, 1745. *See* alſo Gemelli Carreri, *Paris*, 1719. tom. VI. p. 24, &c.

† Willughby and Barrere.

but

but we may ſee them by ſtroking aſide the feathers at the baſe of the bill, which in moſt ſpecies conceal them; in others, they are bare, and conſequently very apparent.

The Toucans have nothing in common with the Woodpeckers, but the diſpoſition of their toes, two placed before and two behind. Even in this character, a diſtinction may be perceived; the toes are much longer and differently proportioned than in Woodpeckers; the outer fore-toe is almoſt as long as the whole foot, which is indeed very ſhort, and the other toes are alſo very long, the two inner ones are the leaſt ſo. The feet of the Toucans are only half the length of the legs, ſo that they cannot walk, but hop, and that awkwardly: the feet are not feathered, but covered with long ſoft ſcales. The nails are proportioned to the length of the toes, arched, and ſomewhat flattened, obtuſe at the end, and furrowed below lengthwiſe by a channel; they are of no ſervice to the bird in attack or defence, or even for climbing, but only to ſupport it firmly on the boughs on which it perches.

The Toucans are ſcattered through all the warm parts of South America, and never occur in the ancient continent. They flit rather than migrate, following the maturity of the fruits on which they feed, particularly

particularly thoſe of the palms; and as theſe trees delight in wet places near the margin of water, the Toucans affect ſuch ſituations, and ſometimes they even lodge on the mangroves, which grow in deluged mud. And hence it has been ſuppoſed that they eat fiſhes *, but theſe muſt at leaſt be very ſmall, ſince they are obliged to ſwallow their food entire.

Theſe birds generally go in ſmall bodies from ſix to ten; their flight is heavy and laborious, owing to the ſhortneſs of their wings and the weight of their enormous bill; yet they riſe above the talleſt trees, on the ſummits of which they are almoſt always perched, and in continual flutter; but the vivacity of their motions diſpels not their dull air, for the huge bill gives them a ſerious melancholy countenance, and their large dull eyes augment the effect: in ſhort, though lively and active, they appear the more heavy and awkward.

As they breed in the holes abandoned by the Woodpeckers, it has been ſuppoſed they excavate theſe themſelves. They lay only two eggs, yet all the ſpecies contain abundance of individuals. They may be eaſily tamed, if taken young; it is even

* Fernandez and Nieremberg.

ſaid

ſaid that they will propagate in the domeſtic ſtate. They are not difficult to rear, for they ſwallow whatever is thrown to them, bread, fleſh, or fiſh; they alſo with the point of their bill, lay hold of the bits that are held near them; they toſs theſe up, and receive them in their wide throat. But when they are obliged to provide for themſelves, and to gather their food from the ground, they ſeem to grope, and ſeize the crumb ſidewiſe, that it may leap up, and be caught in its fall. They are ſo delicate to impreſſions of cold, as to be affected by the coolneſs of evening in the hotteſt climates, even of the new continent; they have been obſerved in the houſe to make a bed of herbs, of ſtraw and of whatever they could gather. Their ſkin is in general blue under the feathers, and their fleſh, though black and hard, is yet palatable.

The genus of theſe birds ſubdivides itſelf into the Toucans and Aracaris: theſe differ from each other; 1. in ſize, the Toucans being much larger than the Aracaris, 2. by the dimenſions and the ſubſtance of their bill, which in the Aracaris is much longer, harder, and ſolider; 3. by their tail, which is longer in the Aracaris, and very ſenſibly tapered, whereas it is rounded in the in the Toucans †. The Toucans properly ſo called, contain five ſpecies.

† The Braſilians were the firſt who diſtinguiſhed theſe two varieties, calling the large ones *toucans*, and the ſmall ones *Aracaris*: The natives of Guiana have made the ſame diſcrimination, giving the former the name of *kararouima*, and the latter that of *grigri*.

The TOCO.

First Species.

Ramphastos-Toco. Gmel.

THE body of this bird is nine inches long, including the head and tail; its bill is seven inches and a half; the head, the upper side of the neck, the back, the rump, the wings, the whole tail, the breast, and the belly, are deep black; the coverts of the upper side of the tail are white, and those of the under side are fine red; the under side of the neck and throat are white, mixed with a little yellow: between this yellow below the throat and the black of the breast, there is a small red circle; the base of the two mandibles is black; the rest of the lower mandible is reddish yellow; the upper mandible is of the same reddish yellow colour, as far as two thirds of its length; the rest of this mandible is black to the point: the wings are short, and reach hardly the third of the tail; the feet and nails are black. This species is new, and we have given it the name of *toco*. [A]

[A] Specific character of the *Ramphastos-Toco*: It is blackish; its throat and rump, white; its orbits, a circle on its breast, and its vent, red.

No. 167

THE TOUCAN.

The YELLOW-THROATED TOUCAN.

Second Species.

{ *Ramphastos Dicolorus.* Linn. and Gmel.
{ *Ramphastos Tucanus.* Linn. and Gmel.

Tucana { *Cayanensis* / *Brasiliensis.* } *gutture luteo.* Briss.

Two birds of this kind have been figured in the *Planches Enluminees*; the first under the denomination of the *Yellow throated Toucan, of Cayenne* [A], and the second under that of the *yellow-throated toucan of Brazil.* [B] But they inhabit equally both countries, and appear to us to form the same species. The difference in the colour of their bill, in the extent of the yellow plate on the throat, as well as in the vivacity of their colours, may be owing to the age of the bird: It is certain, that the superior coverts of the tail are yellow in some individuals, and red in others. In

[A] Specific character of the *Ramphastos Dicolorus.* "It is blackish; its breast, its belly, its vent, and its rump red; its throat yellow."

[B] Specific character of the *Ramphastos-Tucanus.* "It is blackish; a bar on its belly, its vent and ramp yellow."

 both,

both, the head and upper ſide of the body, the wings, and tail, are black ; the throat orange, more or leſs bright ; under the throat, a red bar of various width ſtretches on the breaſt ; the belly is blackiſh, and the inferior coverts of the tail are red ; the bill is black with a blue ſtripe running from its top all its length : the baſe of the bill is ſurrounded by a pretty broad yellow or white bar ; the noſtrils are concealed in the feathers at the root, and their apertures are round ; the legs are twenty lines in length, and bluiſh ; the bill, four inches and a half long, and ſeventeen lines high above the baſe ; the whole bird, from the end of the bill to the extremity of the tail, is nineteen inches ; and if from this we deduct ſix inches and two or three lines for the tail, and four inches and a half for the bill, there remains hardly nine inches for the length of the head and body.

It is from this ſpecies of Toucan, that thoſe brilliant feathers uſed as ornaments are obtained ; all the yellow part is cut off from the ſkin, and ſold at a high price. The males only furniſh theſe fine yellow feathers, for the throat of the females is white ; and this diſtinction has miſled the nomenclators, who have regarded the male and female

female as of different ſpecies *, and finding ſome variation of colours in both, have even gone ſo far as to make each include two ſeparate ſpecies. But we reduce theſe four pretended ſpecies to one, and we may alſo join a fifth mentioned by Laét †, which differs only in the white colour of its breaſt.

In general, the females are very nearly as large as the males; their colours are not ſo vivid, and the red-bar below the throat is very narrow; in other reſpects, they are exactly ſimilar. This ſecond ſpecies is the moſt common, and perhaps the moſt numerous of the Toucans. They abound in Cayenne, particularly in the ſwampy foreſts, and on the mangrove trees. Though like the reſt of the genus, they have only a feathery tongue, they articulate a ſound like *pinien-coin*, which the creoles of Cayenne have employed as its deſignation, but which

* { *Ramphaſtos Piſcivorus.* Linn. and Gmel.
Ramphaſtos Erythrorhynchos. Gmel. }
{ *Tucana Cayanenſis gutture albo.* Briſſ.
Tucana Braſilienſis gutture albo. Briſſ. }
Picus Americanus. Fernandez.
Altera Xochitenacatl. Fernandez.
Paſſer Longiroſtrus, Xochitenacatl dictus. Nieremberg.
{ *The Toucan, or Braſilian Pye.* Edw.
The Redbeaked Toucan. Edw. }

† Hiſtoire du Nouveau Monde. p. 553.

we have not adopted, it being common alſo to the Toco.

Specific character of the *Ramphaſtos Piſcivorus.* "It is blackiſh; the bar on its belly, and its vent, are red; its rump is white."

Specific character of the *Ramphaſtos Erythrorynchos*: "It is blackiſh; its cheeks, and its throat, white; the upper coverts of its tail, brimſtone-coloured; the lower, and a creſcent on its breaſt, red."

The RED-BELLIED TOUCAN.

Third Species.

Ramphaſtos Picatus. Linn. and Gmel.
Tucana. Briſſ.
Naſutus Simpliciter. Klein.
The Braſilian Pye, or Toucan. Alb. and Will.
The Preacher Toucan. Lath.

THIS Toucan has a yellow neck, like the preceding; but its breaſt is a fine red, which in the other is black. Thevet, who firſt mentioned this bird, ſays that its bill is as long as its body. Aldrovandus admits the bill to be two palms in length, and one palm in breadth; and Briſſon reckons the palm three inches. As we have never ſeen the bird, we can only ſpeak from the accounts given by the two former writers. We may obſerve however, that Aldrovandus was miſtaken in aſſigning it three toes before, and one behind; Thevet expreſsly

expreſsly mentions, that it has two before and two behind, which is conformable to nature.

The head, neck, back, and wings, are black, with whitiſh reflections; the breaſt is of a fine gold colour, with red above it, that is, under the throat; the belly, too, and the legs, are of a very bright red, and likewiſe the extremity of the tail, the reſt of which is black; the iris is black, ſurrounded by a white circle, which itſelf is incloſed within another yellow circle; the lower mandible is only half as large near the extremity than the correſponding part of the upper mandible; both are indented on the edges.

Thevet aſſures us, that this bird lives on pepper, of which it ſwallows ſuch quantities as to be obliged to vomit it. This ſtory has been copied by all the naturaliſts; and yet there is no pepper in America. It would be difficult to imagine what ſpice Thevet meant, unleſs it was pimento, which ſome authors have termed *Jamaica pepper*. [A]

[A] Specific character of the *Ramphaſtos Picatus*. "It is blackiſh; its breaſt, yellow; its vent, and the tips of its tail-quills, red; its rump black."

The COCHICAT.

Fourth Species.

Psittacus Torquatus. Gmel.
Ramphastos Torquatus. Lath. Ind.
Tucana Mexicana Torquata. Briss.
The Collared Toucan. Lath. Syn.

WE have shortened the Mexican name *Cochitenacatl* into *Cochicat*. Fernandez is the only author who says he has seen it, and I shall here borrow his account. "It is nearly of the same bulk with the other Toucans; its bill is seven inches long, the upper mandible white and indented, and the lower black; the eyes are black, and the iris reddish yellow: the head and neck, black, as far as a cross red line which encircles it like a collar, after which the upperside of the neck continues black, and the underside whitish, sprinkled with red spots and small black lines; the tail and wings are also black; the belly is green; the legs, red; the feet, of a greenish-ash colour, and the nails black. It frequents the sea-shore, and lives on fish [A]."

[A] Specific character of the *Psittacus Torquatus*: "Above it is black, below, whitish, its belly green, its hind part red, its collar red."

The HOTCHICAT.

Fifth Species.

Ramphaftos Pavoninus. Gmel.
Tucana Mexicana Viridis. Briff.
The Pavonine Toucan. Lath.

THIS name is contracted alfo for the Mexican *Xochitenacatl* *; and Fernandez is the only author who has defcribed it from the life. " It is, he fays, of the fize and fhape of a parrot ; its plumage is almoft entirely green, only fprinkled with fome red fpots ; the legs and feet are black and fhort; the bill is four inches long ; it is variegated with yellow and black." This bird alfo inhabits the fea-coaft in the hotteft parts of Mexico. [A]

* The *Xo* is pronounced *Ho*.

[A] Specific character of the *Ramphaftos Pavoninus*: " It is green, fprinkled with red and iris feathers."

The ARACARIS.

THE Aracaris, as we have already ſaid, are ſmaller than the Toucans; we are acquainted with four ſpecies of them, all natives of America.

The GRIGRI.

Firſt Species of Aracari.

Ramphaſtos Aracari. Linn. and Gmel.
Ramphaſtos Viridis. Linn. and Gmel.

Tucana Braſilienſis Viridis. Briſſ. and Gerini.
Tucana Cayanenſis Viridis. Briſſ.

The Aracari Toucan. Lath.
The Green Toucan. Lath.

THIS bird is found in Braſil, and is very common in Guiana, where it is called *Gri-gri*, becauſe that word expreſſes its ſhrill ſhort cry. It has the ſame natural habits as the Toucans; it alſo inhabits the ſwamps, and lodges among the palms [A]. This ſpecies contains a variety which our nomenclators

[A] Specific character of the *Ramphaſtos-Aracari*: "It is green; the bar on its belly, the vent, and the rump, are red; its belly, bright yellow.

nomenclators have regarded as a ſeparate ſpecies [B]. Yet the difference is ſo ſlight that it may be imputed to age or climate: it conſiſts in a croſs bar of fine red on the breaſt. There is likewiſe ſome difference in the colours of the bill; but that character is quite dubious, ſince theſe vary in the ſame individual according to the age, without following any regular order: ſo that Linnæus was wrong in drawing the ſpecific characters of birds from the colours of the bill.

The head, the throat, and the neck are black; the back, the wings and the tail are dull green; the rump, red; the breaſt and belly, yellow; the inferior coverts of the tail, and the feathers of the legs, olive yellow, variegated with red and fulvous; the eyes large, and the iris yellow; the bill is four inches and a quarter long, ſixteen lines high, and of a more ſolid and harder texture than the bill of the Toucans; the tongue is of the ſame feathery nature; the feet, blackiſh-green, very ſhort, and the toes very long: the whole length of the bird, including the bill and the tail, is ſixteen inches and eight lines.

[B] Specific character of the *Ramphaſtos Viridis*: "It is green, its belly bright yellow, its rump red."

The

The female differs from the male only in the colour of the throat and of the underſide of the neck, which is brown, but black in the male, which uſually has its bill black and white, whereas in the female the lower mandible is black, and the upper one yellow, with a longitudinal black bar reſembling a long narrow feather.

The KOULIK.

Second Species of Aracari.

Ramphaſtos Piperivorus. Linn. and Gmel.
Tucana Cayanenſis Torquata. Briſſ.
The Green Toucan. Edw.
The Piperine Toucan. Lath.

THE word *koulik*, pronounced faſt, reſembles the cry of this bird, which is the reaſon why the Creoles of Cayenne have impoſed that name. It is rather ſmaller than the preceding, and its bill ſhorter in proportion; the head, the throat, the neck, and the breaſt, are black; on the upper ſide of the neck, there is a yellow narrow half-collar; there is a ſpot of the ſame colour on each ſide of the head, behind the eyes; the back, the rump, and the

the wings, are of a fine green, and the belly, which is alſo green, is variegated with blackiſh; the inferior coverts of the tail are reddiſh, but the tail itſelf is green, terminated with red; the feet are blackiſh; the bill is red at the baſe, and black through the reſt; the eyes are encircled by a naked bluiſh membrane.

The female is diſtinguiſhed from the male by the colour of the top of the neck, where the plumage is brown, but black in the male; the underſide of the body, from the throat to the lower part of the belly, is gray in the female, and the half-collar is of a very pale yellow, whereas it is of a fine yellow in the male, and the underſide variegated with different colours. [A]

[A] Specific character of the *Ramphaſtos Piperivorus*: "It is green, its foreſide black, its vent and thighs red."

The BLACK BILLED ARACARI.

Third Species.

Ramphastos Luteus. Gmel.
Tucana Lutea. Briss.
The Black-billed Toucan. Lath.

We derive our account of this bird from Nieremberg. It is as large as a pigeon; its bill is thick, black, and hooked; its eyes, too, are black, but the iris yellow; its wings and tail are variegated with black and white; a black bar rises from the bill, and extends on each side to the breast; the top of the wings is yellow, and the rest of the body yellowish white; the legs and feet are brown, and the nails whitish.

The BLUE ARACARI.

Fourth Species.

Ramphastos Cæruleus. Gmel.
Tucana Cærulea. Briss.
The Blue Tucan. Lath.

No naturalist has seen this bird but Fernandez, and his description is this: "It is of the size of a common pigeon; its bill

bill is very large and indented, yellow above and reddiſh-black below; all its plumage is variegated with cinereous and blue."

It appears from the ſame author, that ſome ſpecies of Aracaris are only birds of paſſage in certain parts of South America.

The BARBICAN.

Bucco Dubius. Gmel.
The Doubtful Barbet. Lath.

I HAVE applied the term *Barbican* to denote a bird that occupies a middle rank between the Barbets and Toucans. It is a new ſpecies, and though it has not hitherto been noticed by naturaliſts, it belongs to no very diſtant climate, for we received it from the coaſt of Barbary, yet without its name, or any account of its natural habits.

The toes are diſpoſed two before and two behind, as in the Barbets and Toucans; it reſembles the latter in the diſtribution of its colours, in the ſhape of its body, and in its large bill, but it is ſhorter, much narrower, and ſolider than that of the Toucans; it reſembles at the ſame time the Barbets in the long briſtles which ſhoot from the baſe of the bill and extend conſiderably beyond the noſtrils; the ſhape of the bill is peculiar, the upper mandible being pointed, hooked at its extremity, with two blunt indentings on each ſide; the lower mandible is ſtriped tranſverſely with ſmall furrows;

rows; the whole bill is reddish, and curved downwards.

The plumage of the Barbican is black on all the upper part of the body, the top of the breast and belly; and it is red on the rest of the under surface, nearly as in that of the Toucans.

It is nine inches long; the tail three inches and a half; the bill eighteen lines long and ten thick, and the legs are only an inch high; so that the bird can hardly walk [A].

[A] Specific character of the *Bucco Dubius*: "It is black, below red; the bar on its breast, and its vent, black."

The CASSICAN.

Coracias Varia. Gmel.
The Pied Roller. Lath.

THIS bird partakes of the properties of the Casiques and Toucans, and therefore I have termed it *Cassican*. It is a new species sent by Sonnerat. We are uncertain what climate it inhabits, only we presume that it comes from the southern parts of America; it certainly resembles the American Cassiques in the shape of its body and the bald spot on the foreside of the head, at the same time that it is analogous to the Toucan in the bulk, and shape of its bill, which is round and broad at the base, and hooked at the extremity. So that if its bill were larger, and the toes disposed two and two, it might be regarded as a proximate species of the Toucans.

It will be superfluous to describe the colours of this bird; its body is slender and long, being about thirteen inches; the bill is two inches and a half; the tail five inches, and the legs fourteen lines. We are unacquainted with its natural habits; but if we were to judge from the shape of its

No. 168

THE CASSICAN.

its bill and feet, we ſhould ſuppoſe that it lives on prey. Yet the Toucans and Parrots which have their bill and nails hooked, ſubſiſt entirely on fruits; and the nails and bill are much leſs hooked in the Caſſican: ſo that we ſhall regard it as a frugivorous bird till we are better informed. [A]

[A] Specific character of the *Coracias Varia*: "It is black; its underſide, the lower part of its back, its rump, and the ſuperior coverts of its tail, white; its tail equal and black, tipt with white." Mr. Latham juſtly obſerves that this bird partakes ſo much of the characters of the Rollers, of the Orioles, and of the Toucans, that it can hardly be referred to and one genus.

The CALAOS, Or RHINOCEROS BIRDS.

WE have feen that the Toucans, fo remarkable for their enormous bill, belong all of them to the continent of South America. We are now to view other birds, natives of Africa and of India, which wear a bill as prodigioufly large, and of a fhape ftill more exceffively monftrous. For nature is more vigorous in the ancient than in the new world, and even her errors are more luxuriant.

When we confider the uncommon expanfion and cumbrous overgrowth which fwells and deforms the bill of thefe birds, we are ftruck with the incongruity and difcordance of their ftructure. But nature exhibits other examples of inconfiftency; the *crofs-bills* are almoft incapacitated for taking their food, and are unable to defend themfelves againft even the weakeft and fmalleft tribes. In the quadrupeds alfo, the floths, the ant-eaters, the fhort-tailed manis, &c. naked or miferable, by reafon of the fhape of their body, and of the difproportion of their members, drag out a laborious and painful painſul exiſtence, continually oppreſſed by the exuberance, or cramped by the deficiency of organization. The life of ſuch feeble imperfect ſpecies is protected by ſolitude alone, and could never be ſupported and maintained, but in deſert receſſes, untrod by man or the powerful animals.

The bill of the Calao, though large, is weak and ill compacted, and ſo far from being uſeful, it proves burthenſome: it is like a long lever where the force is applied near the *fulcrum*, and conſequently the extremity acts feebly: its ſubſtance is ſo tender, that it ſhivers by the leaſt attrition, and theſe accidental cracks have been miſtaken by naturaliſts for a regular and natural indenting. Theſe produce a remarkable effect on the bill of the Rhinoceros Calao; for the mandibles meet only at the point and the reſt remains wide open, as if they were not formed for each other. The interval is worn and broken in ſuch a manner, that this part would ſeem intended to be uſed only at firſt, and afterwards neglected.

We follow our nomenclators in applying the name *calao* to the whole genus of theſe birds, though the people of India have beſtowed it only on one or two ſpecies. Many naturaliſts have given them the appellation

 Rhinoceros,

*Rhinoceros**, on account of the ſort of horn which riſes on the bill; but almoſt all of them have ſeen only the bills of thoſe extraordinary birds †. We will range the calaos according to the moſt ſtriking character, the ſingular ſhape of the bill. We ſhall find that even in her aberrations, nature proceeds by inſenſible gradations, and that of the ten ſpecies which compoſe the genus, there is only one perhaps that merits the deſignation of *Rhinoceros bird.*

These ten ſpecies are;

1. The Calao-rhinoceros.

2. The round-helmeted Calao.

3. The concave helmeted Calao of the Philippines.

4. The Abyſſinian Calao.

5. The African Calao, which we ſhall term *Brac.*

6. The Malabar Calao, which we have ſeen alive.

7. The Molucca Calao, of which we have a ſtuffed preparation.

8 The Calao of the iſland of Panay, of which we have ſtuffed ſpecimens of both the male and female.

* Edwards.—Grew.—Cluſius.—Willughby, &c.

† Edwards, Belon, &c.

9. The

9. The Manilla Calao, of which we have a ſtuffed ſpecimen.

10. Laſtly, the *Tock*, or red-billed Calao of Senegal, of which we have a ſtuffed ſpecimen.

If we invert the order of theſe ten ſpecies, we ſhall be able to trace all the ſteps by which nature arrives at this monſtrous conformation of bill. The Tock has a broad bill faſhioned as the reſt like a ſickle; but it is ſimple, and without any protuberance. In the Manilla Calao, a ſwell may be perceived on the top of the bill: this is more diſtinct in the Molucca Calao: ſtill more conſiderable in the Abyſſinian Calao: the excreſcence is prodigious in thoſe of the Philippines and Malabar, and quite monſtrous in the Rhinoceros-Calao. But though theſe birds admit of ſuch varieties in the bill, they have one general reſemblance in the conformation of the feet, the lateral toes being very long, and almoſt equal to the middle one.

The TOCK.

First Species.

Buceros Nasutus. Linn. and Gmel.

{ *Hydrocorax Senegalensis Erythrorynchos.* } Briss.
{ *Hydrocorax Senegalensis Melanorynchos.* }

The Black-billed Hornbill. Lath.

THIS bird has a very large bill, but this is simple, and without any excrescence; yet still fashioned as in the other Calaos, like a sickle. It resembles them for the most part also in its natural habits, and occurs too in the hottest climates of the ancient continent. The negroes of Senegal give it the name of Tock, which we shall retain. The young bird differs very much from the adult; for its bill is black, and its plumage ash-gray, and, with age, the bill becomes red, and the plumage blackish on the upper side of the body, on the wings, and on the tail, and whitish quite round the head, on the neck, and on all the lower parts of the body: it is also said, that the legs are originally black, and grow afterwards reddish. It is not therefore surprizing that Brisson has made two species; his first designation

ſignation ſeems applicable to the adult Tock, and his ſecond to the young Tock.

This bird has three toes before and only one behind: the middle one is cloſely connected to the outer toe as far as the third joint, and much more looſely to the inner toe, and at the firſt joint only. The bill is very thick, bent downwards, and ſlightly indented on the ſides.

The ſubject which we deſcribe was twenty inches long; its tail ſix inches, ten lines: its bill three inches five lines, and twelve lines and a half thick at the baſe; its horny ſubſtance of the bill is thin and light: the legs are eighteen lines high.

Theſe birds which are pretty common in Senegal, are very ſimple when young; they ſuffer a perſon to approach and catch them; and they may be ſhot at without being ſcared, or even without their ſtirring. But age inſtructs them by experience; their diſpoſition quite alters; they grow extremely ſhy, and eſcape to the ſummits of the trees; while the young ones remain on the loweſt branches and the buſhes, and continue perfectly ſtill, their head ſunk between the ſhoulders, ſo that the bill only is ſeen. The young birds ſcarce ever fly, whereas the old ones ſoar in a lofty and rapid courſe.

The young Tocks are numerous in the months of Auguſt and September: they may be caught with the hand, and appear as tame as if they had been reared in the houſe. But this conduct proceeds from their ſtupidity, for they do not pick up the food that is thrown to them, and it muſt be put in their bill; which affords a preſumption that the parents are obliged to maintain them for a great length of time.

The Tock is very different from the Toucan, and yet one of our moſt intelligent naturaliſts has confounded them. Adanſon ſays, in his voyage to Senegal, that he killed two Toucans in that country; but it is certain that there are no Toucans in Africa. I ſhould therefore preſume that our philoſophical traveller meant the Tocks. [A]

[A] Specific character of the *Buceros Naſutus*: "Its front is ſmooth, its tail-quills white at the baſe and at the tip."

The MANILLA CALAO.

Second Species.

Buceros Manillensis. Gmel.
The Manilla Hornbill. Lath.

THIS species was hitherto unknown; it was sent to the King's Cabinet by M. Poivre, to whom we owe much other information, and many curious preparations. The bird is scarcely larger than the Tock; it is 20 inches long; its bill two inches and a half, less curved than that of the Tock, not indented, but as sharp at the edges, and more pointed; above this bill, there is a prominent light festoon, adhering to the upper mandible, and forming a simple inflation; the head and neck are white, stained with yellowish, and marked with brown waves; there is a black spot on each side of the head at the ears; the upperside of the body is blackish brown, with some whitish fringes wrought slightly in the quills of the wing; the underside of the body is dirty white; the quills of the tail are of the same colour with those of the wings, only they are intersected in their middle by a rufous bar,

bar, two fingers broad. We are unacquainted with the œconomy of this bird. [A]

[A] Specific character of the *Buceros Manilliensis*: "Above it is blackish brown, belcw dirty white, its bill not serrated; a small prominence."

The CALAO, of the Island of Panay.

Third Species.

Buceros Panayensis. Gmel.
The Panayan Hornbill. Lath.

THIS bird was brought by M. Sonnerat, Correspondent of the Cabinet. I shall transcribe the account which he gives of it in his voyage to New Guinea: he calls it the *chiseled bill Calao*; but that epithet is equally applicable to other Calaos.

"The male and female are of the same size, and nearly as large as the great European Raven, rather longer and narrower shaped; their bill is very long and arched into the form of a sickle, indented along its edges above and below, terminated by a sharp point, and depressed at its sides; it is furrowed up and down, or across two thirds of its length; the convex part of these furrows is brown, and the concave spaces are of the colour of orpiment; the rest of the bill

bill near the point, is thin and brown; at the root of the bill, there rises upwards an excrescence of the same horny substance, flat at the sides, sharp above, and cut at right angles before; this excrescence extends along the bill to its middle, where it terminates, and its uniform height is equal to half the breadth of the bill; the eye is encircled by a brown membrane devoid of feathers; the eyelid bears a ring of hard, short, stiff bristles, which form real eyelashes; the iris is whitish: in the male, the head, the neck, the back, and the wings are greenish black, changing into bluish, according to the position. In the female, the head and the neck are white, except a broad triangular spot, which extends from the base of the bill, below and behind the eye, as far as the middle of the neck across the sides; this spot is dark green, fluctuating like the neck and back in the male: the female has also the neck and wings of the same colour as in the male; the top of the breast in both sexes is of a light brown red; the belly, the thighs, and the rump are equally of a deep brown red; they both have ten quills in the tail, of which the upper two thirds are of a rusty yellow, and the lower third is a black transverse bar; the feet are lead colour, and

composed

composed of four toes, one of which is directed behind and three before; the middle one is connected to the outer toe as far as the third joint, and to the inner toe as far as the first only." [A]

[A] Specific character of the *Buceros Panayensis*. "It is greenish black, below red brown, the prominence of the upper mandible sharp above, and flat at the sides."

The MOLUCCA CALAO.

Fourth Species.

Buceros-Hydrocorax. Linn. and Gmel.
Hydrocorax Briss.
Corvus Indicus Bontii. Will.
Corvus Torquatus. Klein.
Cariocatactes. Mœrh.
The Indian Hornbill. Lath.

THE name *Alcatraz* has improperly been applied to this bird; and Clusius is the author of the mistake. He has inaccurately translated the passage of Oviedo; for the Spanish word *alcatraz*, according to Fernandez, Hernandez, and Nieremberg, denotes the Pelican of Mexico. This mistake has occasioned another, which our nomenclators have transferred to the whole genus of the Calaos; they suppose that these birds

birds haunt the margin of water, and hence they bestow the appellation of *hydrocorax* (Water-raven.) But this opinion is refuted by all the obsservers who have viewed them in their native abodes: Bontius, Camel, and, what carries still more weight, the Calao itself, by its structure, the shape of its feet and bill, demonstrate that it is neither a Raven nor an aquatic bird.

The Molucca Calao is two feet four inches long; the bill eight inches; but the legs are only two inches two lines: this character of the shortness of the legs belongs not only to it, but to all the other Calaos, which walk with the utmost difficulty; the bill is five inches long, and two inches and a half thick at its origin; it is blackish cinereous, and supports an excrescence, whose substance is solid and like horn; this excrescence is flat before, and extends rounded to the upperside of the head; it has large black eyes, and its aspect is disagreeable; the sides of the head, the wings, and the throat, are black, and that part of the throat is surrounded with a white bar; the quills of the tail are whitish gray; all the rest of the plumage is variegated with brown, with gray, with blackish, and with fulvous; the feet are brown gray, and the bill blackish.

These

These birds, says Bontius, do not live on flesh, but on fruits and particularly nutmegs, to which they prove very destructful; and that food communicates to the flesh, which is tender and delicate, an aromatic odour, that renders it more grateful to the palate. [A]

[A] Specific character of the *Buceros Hydrocorax*: "Its bony front is flat and truncated before, its belly fulvous."

The MALABAR CALAO.

Fifth Species.

Buceros Malabaricus. Gmel.
The Pied Hornbill. Lath.

THIS bird was brought from Pondichery; it lived the whole of the summer 1777, in the court-yard of the Marchioness de Pons, who was so kind as to present it to me, for which I take this opportunity of expressing the warmest gratitude. It was as large as a Raven, or twice as large as the common crow; it was two feet and a half long, from the point of the bill to the extremity of the tail, which had dropt off in its passage home, and the feathers were begun to grow again, but had

No. 169

THE HORN-BILL.

had not nearly attained their full ſize: ſo that we may preſume that the entire length of the bird was about three feet; its bill was eight inches long and two broad, and bent fifteen lines from the ſtraight poſition; a ſecond bill, if it may be ſo called, ſat like a horn cloſe on the firſt, and followed its curvature, and extended from the baſe to within two inches of the point of the bill, it roſe two inches three lines; ſo that, meaſuring in the middle, the bill together with its horn formed a height of four inches; near the head, they were both of them fifteen lines thick acroſs; the horn was ſix inches long, and its extremity appeared to have been ſhortened and ſplit by accident, ſo that we may reckon it to be half an inch longer: on the whole, this horn has the ſhape of a true bill, truncated and cloſed at the extremity, but the junction is marked by a very perceptible furrow, drawn near the middle and following all the curvature of this falſe bill, which does not adhere to the ſkull; but its poſterior portion, which riſes on the head, is ſtill more extraordinary, it is naked and fleſhy, and covered with quick ſkin, through which this paraſite member receives the nutritious juices.

The

The true bill terminates in a blunt point; it is ſtrong and conſiſting of a horny, and almoſt bony, ſubſtance, extended in *laminæ*, and we may perceive the layers and undulations: the falſe bill is much thinner, and may be bent even by the fingers; it is of a light ſubſtance diſpoſed internally in little cells, which Edwards compares to thoſe of a honey comb: Wormius ſays that this falſe bill conſiſts of a matter like crabs ſhells.

The falſe bill is black, from the point to three inches behind, and there is a line of the ſame black at its origin, and alſo at the root of the true bill; all the reſt is yellowiſh-white. Wormius remarks preciſely the ſame colours; only he adds that the inſide of the bill and of the palate, is black.

A white folded ſkin meets the root of the true bill above on both ſides, and is inſerted near the corners of the bill in the black ſkin that encircles the eyes; the eyelid is bordered with long laſhes arched behind; the eye is red brown, and grows animated and ſparkles when the bird is in commotion; the head, which appears ſmall in proportion to the enormous bill which it bears, reſembles much in its ſhape that of the Jay: in general, the figure, the geſture, and the whole form of this Calao, ap-

appears to us composed of the features and movements of the Jay, the Raven, and the Magpie. These resemblances have struck most observers, and hence the bird has been called, *the India Raven*, *the Horned Crow*, *The Horned Pie of Ethiopia.*

The feathers on the head and neck were black, which it had the power of bristling, and they were often in that state, as in the Jay; those of the back and wings were also black, and all of them had a slight reflection of violet and green: on some of the coverts of the wings, there was an edging irregularly traced, and the feathers seemed bunched out like those of the Jay; the stomach and belly were of a dirty white; among the great quills of the wings, which are black, the outer ones only are white at the point; the tail, which had begun to grow again, consisted of six white quills black at the root, and four which were springing from their shafts entirely black; the legs are black, thick, and strong, and covered with broad scales; the nails, which were long, but not sharp, seemed calculated for holding and clenching. This bird hopped with both its feet at once, forward and sidewise, like the Jay and Magpie, but did not walk. When at rest, its head seemed to

 recline

recline on its ſhoulders; when diſturbed by ſurprize and inquietude, it ſwelled and raiſed itſelf, and ſeemed to aſſume an air of boldneſs and importance. Yet its uſual gait was mean and ſtupid; its movements ſudden and diſagreeable; and its reſemblance in features to the Magpie and the Raven, gave it an ignoble aſpect *, ſuited to its diſpoſitions. Though there are ſpecies of Calaos that appear to be frugivorus, and we have ſeen this bird eat lettuces, which it firſt bruiſed in its bill; it ſwallowed raw fleſh: it caught rats, and devoured even a ſmall bird, which was thrown to it alive. It often repeated the hoarſe cry *ook*, *ook*; it uttered alſo from time to time another ſound, which was feebler, and not ſo raucous, and exactly like the clucking of a Turkey-hen when ſhe leads her brood.

We have ſeen it ſpread and open its wings to the ſun, and ſhudder at a paſſing cloud, or a ſlight breeze. It did not live more than three months at Paris, and died before the end of ſummer. Our climate is therefore too cold for it.

We cannot forbear remarking, that Briſſon is miſtaken in referring the figure *d* of the bill in *Pl.* 281 of Edwards Gleanings to his Philippine Calao: for that

* Bontius.

figure

figure repreſents our Malabar Calao, which bears a ſimple excreſcence, and not a concave double horned-helmet, like the Philippine Calao. [A]

[A] Specific character of the *Bucco Malabaricus*: "It is black, below white; the prominence of its front rounded above, ſharp towards the front, extended behind the eyes."

The BRAC or AFRICAN CALAO.

Sixth Species.

Buceros Africanus. Gmel.
Hydrocorax Africanus. Briſſ.
The African Hornbill. Lath.

We ſhall retain the name *Brac* given to this Calao by father Labat, eſpecially as that traveller is the only perſon that has ſeen and obſerved it. It is very large, its head alone and its bill making together eighteen inches in length; this bill is partly yellow, and partly red; the two mandibles are edged with black: at the upper part of the bill, there is an excreſcence of a horny ſubſtance, which is of the ſame colour, and of a conſiderable ſize; the fore part of this excreſcence projects forward like a horn, it is almoſt ſtraight, and does not bend

 upwards;

upwards; the hind part is on the contrary rounded and covers the top of the head; the nostrils are placed below this excrescence, near the origin of the bill: the plumage of this Calao is entirely black.

The ABYSSINIAN CALAO.

Seventh Species.

Buceros Abyssinicus. Gmel.
The Abyssinian Hornbill. Lath.

THIS Calao appears to be one of the largest of the genus; yet, if we were to judge from the length and thickness of the bills, we should reckon the Rhinoceros Calao still larger. The bill of the Abyssinian Calao seems fashioned after that of the Raven, only it is more bulky; the total length of the bird is three feet two inches; it is entirely black, except the great quills of the wings, which are white, the middle ones and a part of the coverts, which are of a deep tawny brown; the bill has an easy, equal arch through its whole length, and it is flat and compressed at the sides; the two mandibles are hallowed internally with furrows, and terminate in a blunt point. The bill is nine inches long; it bears a semicircular

circular prominence that reaches from its root to near the front, two inches and a half in diameter, and fifteen lines broad at its base over the eyes; this excrescence is of the same substance with the bill, but thinner and yields under the fingers; the height of the bill taken vertically and joined to that of its horn, is three inches eight lines; the feet measure five inches and a half; the great toe, including the nail, is twenty-eight lines; the three fore toes are almost equal; the hind one is also very long, being two inches; all of them are thick, and covered, as well as the legs, with blackish scales, and furnished with strong nails, but which are neither hooked nor sharp: on each side of the upper mandible near its origin, is a reddish spot; the eyelids are provided with long lashes: a naked skin of violet brown encircles the eyes, and covers the throat and part of the foreside of the neck.

The PHILIPPINE CALAO.

Eighth Species.

Buceros Bicornis. Linn. and Gmel.
Hydrocorax Philippensis. Briss.
The Philippine Hornbill. Lath.

THIS bird is, according to Brisson, of the size of the Turkey-hen; but its head is proportionally much larger, which is indeed requisite to support a bill nine inches long, and two inches eight lines thick, and which carries, above the upper mandible, a horny excrescence six inches long, and three inches broad; this excrescence is a little concave on the upper part, and the two anterior angles are produced before into the shape of a double horn; it extends rounding on the upper part of the head; the nostrils are placed near the origin of the bill, below this excrescence. All the bill, as well as this excrescence, is of a reddish colour.

The head, the throat, the neck, the upperside of the body, and the superior coverts of the wings and tail, are black; all the underside of the body is white; the quills of the wings are black, and marked with a white a white ſpot; all the quills of the tail are entirely black, except the two exterior ones, which are white; the legs are greeniſh.

George Camel has, along with the other birds of the Philippines, deſcribed a ſpecies of Calao, which ſeems to be much like the preſent, but not exactly the ſame. His account was communicated to the Royal Society of London, by Dr. Petiver, and printed in the Philoſophical Tranſactions, *No.* 285. *Art. III.* It there appears, that this bird termed *Calao* or *Cagao* by the people of India, does not haunt ſtreams, but inhabits the uplands, and even the mountains, living on the fruits of the *baliti*, which is a ſort of wild fig-tree, and alſo on almonds, piſtachio nuts, &c. which it ſwallows entire. We ſhall here inſert a tranſlation: "The bill is black; the rump and back duſky aſh colour; the head ſmall, and black about the eyes; the eye laſhes black and long; the eyes blue; the bill is a ſpan in length; ſomewhat curved, ſerrated, diaphanous, and of the colour of cinnabar; the ower mandible is about an inch and a half broad at the middle; the upper mandible is a palm in height, more than a ſpan in length, the top flat and about a ſpan in meaſure, and bearing a helmet of a palm in breadth. The tongue is ſmall for ſuch a bill, being hardly

an

an inch. Its voice resembles more the grunting of a sow, or the bellowing of an ox, than the cry of a bird. Its legs with the thighs are a span in length: the feet have three toes before, and only one behind, which are scaly, reddish, and armed with solid black talons: the tail consists of eight great white quills, of a cubit in length; the quills of the wings are yellow. The Gentiles revere this bird, and relate stories of its fighting with the Crane which they call *Tipul* or *Tibol*; they say, that after this battle, the Cranes were obliged to remain in the wet grounds, and that the Calaos would not suffer them to approach the mountains *."

* We are sorry to remark that the translation which the Count de Buffon here gives is exceedingly inaccurate. *Sesquiuncia* is rendered *half an inch*, &c. We have therefore altered it in some places; but, as the last sentence is that from which our ingenious author draws his conclusion, we have preserved it as it stood in the text. We shall now compare it with the original: "*Calao*, (says Camel,) Gentiles superstitiose colunt et observant, fabulantur cum Grue *Tipul* seu Tihol pactasse, ut hæc palustribus, Calao sylvosis, contenta viverent (hinc Tipol si ligno quocunque insederit, in pænam transgressi fœderis sese loco movere non valere, e contra Calao si aquosis et humilibus." That is, the idolatrous Indians have a superstitious veneration for the Calao, and relate that it has entered into a compact with the Crane that it should live contented with its marshes, and the Calao with the woods, thence the Crane, if it perch on a tree, cannot stir from the spot, as a punishment for infringing the treaty; and on the other hand, the Calao incurs the same punishment, if it alights in the low fens.

This

This ſort of deſcription ſeems to prove clearly that the Calaos are not aquatic birds; and as the colours and ſome other characters are different from thoſe of the Philippine Calao, deſcribed by Briſſon, we conceive that this ſhould be reckoned a variety of the other.

The ROUND HELMETED CALAO.

Ninth Species.

Buceros Galeatus.

The Helmet Hornbill. Lath.

WE have only the bill of this bird, and it is like that given by Edwards. If we judge of the ſize of the bird from the bulk of the head, which remains attached to the bill, this Calao is one of the largeſt and ſtrongeſt of the genus; the bill is ſix inches long, from the corners to the point; it is almoſt ſtraight, and not indented: from the middle of the upper mandible there riſes and extends as far as the occiput, a wen ſhaped like a helmet, two inches high, almoſt round, but a little compreſſed on the

the ſides; this protuberance where it joins the bill, has an altitude of four inches and a circumference of eight. The faded and embrowned colours of this bill, which is depoſited in the Cabinet, no longer exhibit that vermillion tinge which appears in Edwards' figure.

Aldrovandus gives a diſtinct figure of the bill of this round helmeted Calao under the name of *Semenda, a bird of India, whoſe hiſtory is ſtill almoſt entirely fabulous.* This bill, which belonged to the Cabinet of the Grand Duke of Tuſcany, had been brought from Damaſcus. The helmet was of an oval ſhape; it was white before and red behind; the bill was a palm in length, pointed, and channeled. When we compare this deſcription with the figure, we find that this is the bill of the round helmeted Calao.

The RHINOCEROS CALAO.

Tenth Species.

Buceros Rhinoceros. Linn. and Gmel.
Nasutus Rhinoceros. Klein.
Hydrocorax Indicus. Briss.
Topau. Borousk.
Rhinoceros Avis. Johnst.
Corvus Indicus Cornutus. Bontius.
Tragopan. Moehring.
The Horned Pie of Ethiopia. Charlton.
The Horned Indian Raven, or Rhinoceros Bird. Will. & Edw.
The Rhinoceros Hornbill. Lath.

SOME Authors have confounded this bird with the *Tragopanda* of Pliny, which is the cassowary, known to the Greeks and Romans, and which was found in Barbary and the Levant, very remote from the native seat of this Calao.

The Rhinoceros bird seen by Bontius in the island of Java, is much larger than the European Raven; it has an offensive smell, and is very ugly. Bontius thus proceeds; "Its plumage is all black, and its bill oddly fashioned; for on its upper part, there rises an excrescence of a horny substance which extends forward, and then bends back towards the top like a horn, and which is is of a prodigious ſize, for it is eight inches long, and four broad at the baſe: this horn is variegated with red and yellow, and as if divided into two parts by a black line that extends on each of the ſides lengthwiſe; the noſtrils are placed below this excreſcence near the root of the bill. It is found in Sumatra, in the Philippines, and in other hot parts of India."

Bontius relates ſome particulars with regard to the œconomy of theſe birds; he ſays that they live on carrion, and commonly follow the hunters of wild cows, boars, and ſtags, who to leſſen the trouble of conveying their game, are obliged to divide carcaſes, and ſend them to the boats on the river, leſt the Calaos ſhould devour them. Yet theſe birds attack no animals but rats and mice, and for that reaſon the Indians rear ſome of them. Bontius tells us that the Calao firſt flattens the mouſe in its bill to ſoften it; and then toſſes it in the air, receives it in its wide throat, and ſwallows it entire: indeed this is the only mode of eating compatible with the ſtructure of its bill, and the ſmallneſs of its tongue, which is concealed almoſt in the throat †.

† Philoſophical Tranſactions, *No.* 285.

Such

Such is the manner of life to which nature has reduced it, by beſtowing a bill ſtrong enough for prey, but too weak for fighting; cumbrous for uſe, a mere ſhapeleſs exuberance. The external ſuperfluities and defects ſeem to have affected the mental faculties of the animal: it is melancholy and ſavage; its aſpect is coarſe, its attitude heavy.—Bontius' figure is inaccurate.

The KING-FISHER.

Le Martin Pecheur, ou L'Alcyon. Buff.
* *Alcedo Iſpida.* Linn. Gmel.
Iſpida. Briſſ. Aldrov. Will. Klein, &c.
Alcedo Fluviatilis. Schwenckf.

THE French name *Martin-fiſher* is beſtowed on this bird, becauſe, like the Martin, it ſkims along the ſurface. Its ancient appellation *Alcyon* or *Halcyon* is nobler. It was celebrated among the Greeks: the epithet *Alcyonian* was applied by them to the four days before and after the winter ſolſtice †, when the ſky is ſerene, and the ſea ſmooth and tranquil ‡. Then,

* In Greek Αλκυων, Κηυξ and Κηρυλος; in modern Greek Φασιδυνις: in Latin *Alcedo and Alcyon*; in modern Latin *Iſpida*; in Italian *Piombino, Picupiolo & Uccello Peſcatore, Uccello del Paradiſo, Uccello della Madonna, Peſcatore del Re*; (i. e. Fiſher-bird, Bird of Paradiſe, Bird of our Lady, King-fiſher); in Lombardy *Merlo Acquarolo,* (i. e. water Black-bird); in German, *Eiſs-vogel, Waſſer heunlein,* and *See Schwalme* (Ice-bird, Water-pullet, Sea-ſwallow); in Poliſh *Zimorodek Rzeczny.*

† Seven according to Ariſtotle.

‡ We ſhall quote Ovid's deſcription:

Perque dies placidos hiberno tempore ſeptem
Incubat halcyone pendentibus æquore nidis:
Tum via tuta maris: ventos cuſtodit, et arcet
Æolus egreſſu. Met. *lib.* xi.

Thus

Nº. 170

THE KING FISHEF

Then, the timorous navigators of antiquity ventured to lose sight of the shore, and to shape their course on the glassy surface of the main. This kind suspension of the horrors of the season, this happy interval of calm was granted the Alcyon, to breed her young ‡. Imagination amplified the simple beauties of nature by the addition of the marvellous; the nest of that bird was made to float on the placid face of the ocean ||; Æolus bound up his winds; Alcyone, his plaintive solitary daughter §, still called on the billows to restore her hapless Ceyx, whom Neptune had drowned, &c. **

This mythological tale of the bird Alcyon is, like every other fable, only the emblem of its natural history; and it is astonishing that Aldrovandus should close his long discussion on the subject by concluding that the bird is now unknown. The description of Aristotle alone sufficiently discriminates it, and shews that it is the same with the

Thus translated by Mr. Dryden.

" Alcyone compress'd
" Seven days sits brooding on her watery nest
" A wintry queen; her sire at length is kind,
" Calms every storm, and hushes every wind. T.

‡ Aristotle, *Hist. Anim.* lib. v. 8.

|| Ælian and Plutarch.

§ " Desertus alloquor alcyonas." *Propertius.*

** Euripides, Ovid, Ariosto.

King-fisher

King fiſher. *The Alcyon*, ſays that philoſopher, *is not much larger than a ſparrow; its plumage is painted with blue and green, and lightly tinged with purple, theſe colours are not diſtinct, but melted together, and ſhining variouſly over the whole body, the wings, and the neck; its bill is yellowiſh* *, *long and ſlender* †.

It is equally characterized by the compariſon of its natural habits. The Alcyon was ſolitary and penſive; the King-fiſher is ſeen almoſt always alone, and the pairing ſeaſon is of ſhort duration ||. Ariſtotle, while he repreſents the Alcyon as an inhabitant of the ſea-ſhore, relates that it alſo aſcends high up the rivers, and haunts their banks: and there is no reaſon to doubt, but that the river King-fiſher is equally fond of the ſea-ſhore, where it can obtain every convenience ſuited to its mode of life. The fact is proved by eye-witneſſes †; yet Klein denies it, though he ſpeaks only

* The epithet υποχλωρος is tranſlated *greeniſh* by Gaza; Scaliger more properly renders it *yellowiſh*. The primitive χλοη is applied not only to the verdant mead, but to the yellow harveſt. T

† Lib. ix. 14.

|| Aldrovandus.

† The King fiſher is fond of the brink of the ſea and of the little rivulets which flow into it; it feeds on the ſmalleſt ſhell-fiſh, takes them in its bill, and cruſhes them by daſhing againſt the pebbles. It ſeeks alſo the large worms which abound on the ſea-ſhore. Its fleſh ſmells of muſk. *Note accompanying a package from M. Guys.*

of

of the Baltic Sea, and was little acquainted, as we ſhall find, with the King-fiſher. The Alcyon was not common in Greece or Italy; Chœrephon, in Lucian's Dialogue §, admires its ſong as a novelty. Ariſtotle and Pliny ſay, that the appearance of the Alcyon was rare and fugitive, that it wheeled rapidly round the ſhips, and inſtantly retired into its little grot on the ſhore ‡. This character agrees perfectly with the King-fiſher, which is ſeldom ſeen.

The King-fiſher may be recognized alſo from the mode in which the Alcyon caught its prey: Lycophron calls it the diver *, and Oppian ſays that *it drops perpendicularly, and plunges into the Sea.* This peculiarity of diving vertically has given occaſion to the Italian appellation *piombino*, or plummet. Thus all the external characters, and all the natural habits of our King-fiſher, are applicable to the Alcyon deſcribed by Ariſtotle. The poets repreſented the neſt of the Alcyon as floating on the ſurface of the ſea: the naturaliſts have diſcovered that it has no neſt, but drops its eggs in horizontal holes in the brinks of rivers or in the ſea-beach.

§ *Dial. Alcyon.*

‡ Ariſtotle, lib. v. 9. Pliny, lib. v. 9.

* Δυπτη

 The

The love ſeaſon of the Alcyon, which was placed about the time of the ſolſtice, is the only circumſtance that does not coincide with the hiſtory of the King fiſher, which however breeds early and before the vernal equinox. But, beſides that this fable may have been added for embelliſhment, it is poſſible that, in a hotter climate, the amours of the King-fiſher commenced earlier. There are different opinions too with regard to the time of the Alcyonian days. Ariſtotle ſays that in the Grecian Archipelago, they were not always contiguous to the ſolſtice, but more conſtantly ſo in the ſea of Sicily †. Nor did the ancients agree in reſpect to the number of theſe days ||. And Columella refers them to the Kalends of March ‡ which is the time when our King-fiſher begins to hatch.

Ariſtotle ſpeaks diſtinctly of one kind only of Alcyon; and it is from a doubtful and probably corrupted paſſage, where according to the correction of Geſner, he treats of two ſpecies of Swallows *, that natura-

† *Lib. v.* 8.

|| *Vide* Coel. Rhodig, *lect. antiq.* lib. xiv. 11.

‡ *Ibidem.*

Lib. VIII. 3. *Το των Αηδονων γενος*, which Gaza and Niphus tranſlate *Alcedones*, though *αηδων* properly ſignifies a nightingale. It were better therefore to read with Geſner, *Χελιδονων*, or "*the tribe of ſwallows*;" eſpecially as in the following line, Ariſtotle begins to ſpeak diſtinctly of the Alcyon as of a different bird.

liſts

lifts have inferred the exiftence of two Alcyons, the one fmall and endowed with voice, the other large and dumb. Belon makes the need thrufh the *vocal Alcyon*, and the King-fifher the *mute Alcyon*, though its character is quite the reverfe.

Thefe critical difcuffions feemed neceffary in a fubject which moft of the naturalifts have left in the greateft obfcurity. Klein, who makes the fame remark, only increafes the confufion, by afcribing to the King-fifher two toes before and two behind. He appeals to Schwenckfield, who has fallen into the fame error †, and to a bad figure of Belon, which however that naturalift has himfelf corrected, by defcribing accurately the foot of this bird, which is fingular. Of the three fore toes, the outer one is clofely connected to that of the middle as far as the third joint, fo as apparently to make only one toe, which forms below a broad flat fole; the inner toe is very fhort, more fo than the hind one; the legs are alfo very fhort; the head is large; the bill long, thick at the bafe, tapered ftraight to a point; the tail is commonly fhort in this genus.

It is the handfomeft bird in our climates; none in Europe can compare with the King-

† This error was firft propagated by Albertus, as Aldrovandus remarks, while he rectifies it.

 fifher

fisher in the elegance, the richness, and the luxuriance of colours; it has all the shades of the rainbow, the brilliancy of enamel, and the glossy softness of silk: all the middle of the back, with the upper surface of the tail, is light blue and brilliant, which, in the sun has the play of sapphire, and the lustre of turquois stone; green is mixed on the wings with blue, and most of the feathers are terminated and dotted with the tints of beryl; the head and the upper side of the neck are dotted in the same manner, with lighter specks on an azure ground. Gesner compares the glowing yellow-red, which colours the breast, to the red glare of a burning coal.

It would seem that the King-fisher has strayed from those climates where the sun pours incessant torrents of the purest light, and sheds all the treasures of the richest colours *. And though our species belongs not precisely to the countries of the East and South; yet the whole genus of these charming birds inhabits those genial regions.

* "There is a species of King-fisher common in all the islands of the South Sea: we have remarked that its plumage is much more brilliant between the tropics, than in the regions situated beyond the temperate zone, as in New Zealand." Forster, *Observations made in Captain Cook's second Voyage.* The King-fisher is called *eroore*, in the language of the Society Islands.

There

There are twenty ſpecies in Africa and in Aſia, and we are acquainted with eight more, that are ſettled in the warm parts of America. Even the European King-fiſher is ſcattered through Aſia and Africa; for many King fiſhers ſent from China and Egypt are found to be the ſame with ours, and Belon ſays that he met with them in Greece †, and in Thrace ‡.

This bird, though it derives its origin from the hotteſt climates, is reconciled to the rigours of our ſeaſons. It is ſeen in the winter along the brooks, diving under the ice and emerging with its fiſhy-prey ||. Hence the Germans have called it *Eiſſvogel*, or Ice-bird §; and Belon is miſtaken in aſſerting that it is only migratory in our climates.

It ſpins with a rapid flight; it uſually traces the windings of the rivulets, razing the ſurface of the water. It ſcreams while on the wing *ki*, *ki*, *ki*, *ki*, with a ſhrill voice, which makes the banks to reſound.

† Nat. des Oiſeaux, p. 220.

‡ The banks of the river Hebrus, (now Meliſſa) are in ſome places pretty high, where the river-alcyons, vulgarly called the *martinets pêcheurs* (King-fiſhers), make their neſts." Id. *Obſervations*, p. 63. The King fiſher is probably not found in Sweden, ſince Linnæus does not mention it: but we are more ſurprized that he places in that northern climate the *Bee-eater*, which is little known in France, and even rare in Italy.

|| Schwenckfeld, Geſner, Olina.

§ Geſner.

In

In ſpring, it has another ſong which may be heard through the murmur of the ſtream, or the noiſy daſhing of the caſcade ‡. It is very ſhy, and eſcapes to a diſtance; it ſits on a branch projecting over the current; there it remains motionleſs, and often watches whole hours to catch the moment when a little fiſh ſprings under its ſtation; it dives perpendicularly into the water, where it continues ſeveral ſeconds, and then brings up the fiſh; which it carries to land, beats to death, and then ſwallows it.

When the King-fiſher cannot find a projecting bough, it ſits on ſome ſtone near the brink, or even on the gravel; but the moment it perceives the little fiſh, it takes a ſpring of twelve or fifteen feet and drops perpendicularly from that height. Often it is obſerved to ſtop ſhort in its rapid courſe and remain ſtill, hovering in the ſame ſpot for ſeveral ſeconds: ſuch is its manner in winter when the muddy ſwell of the ſtream, or the thickneſs of the ice, conſtrains it to leave the rivers, and to ply along the ſides of the unfrozen brooks: at each pauſe, it continues as it were ſuſpended at the height of fifteen or twenty feet;

‡ The name *Iſpida* is, according to the author *de natura rerum*, in Geſner, formed from the cry of the bird; probably from the firſt.

and

and when it would change its place, it ſinks and ſkims along within a foot of the ſurface of the water; then riſes and halts again. This repeated and almoſt continual exerciſe ſhows that the bird dives for many ſmall objects, fiſhes or inſects, and often in vain; for, in this way, it travels many a league.

It neſtles in the brinks of rivers and brooks, in holes made by the water rats, or by crabs, which it deepens and faſhions; and contracts the aperture. Small fiſh-bones and ſcales are found in it among ſand, but without any arrangement, and here its eggs are depoſited: though we cannot find thoſe little pellets with which Belon ſays it plaſters its neſt, or trace the form imputed to it by Ariſtotle, who compares this neſt to a gourd, and its ſubſtance and texture to thoſe ſea-balls or lumps of interwoven filaments which cut with difficulty, but when dried become friable.* The *halcyonia*, of which Pliny reckons four kinds †, and which ſome have ſuppoſed to be the neſts of the King-fiſhers, are only cluſters of Sea-weeds: and with regard to the famous neſts of Tonquin and China, which are eſteemed ſuch delicacies, and have alſo been aſcribed to the Alcyon, we demonſtrated that they

* Αλος-αχνη, or Sea-ſpume. *Hiſt. Anim.* Lib. ix. 14.
† Lib. xxxii. 8.

were the production of the Salangane or esculent Swallow

The King fishers begin to frequent their hole in the month of March; and at this time the male is observed in ardent pursuit of the female. The ancients believed that the Alcyons were extremely amorous, for they relate that the male expires in the embrace *; and Aristotle asserts that they begin to breed when only four months old †.

The species of our King-fisher is not numerous, though these birds have six, seven, or even nine in a hatch, according to Gesner. Their mode of life proves often fatal, nor do they always with impunity brave the rigours of our winters, for they are found dead under the ice. Olina describes the method of catching them at day break, or in the dusk of the evening by setting a trap at the edge of the water; he adds, that they live four or five years. We only know that they may be kept some time in rooms, where are placed basons of water full of small fish ‡. M Daubenton, of the Academy of Sciences, fed some for several months by giving them every day young fry, which is

* Tzetzes, and the Scholiast of Aristophanes.

† Lib. ix. 14.

‡ Vosmaér.

their

their only proper nouriſhment: for four King fiſhers were brought to me on the twenty firſt of Auguſt, 1778; which were as large as their parents, though taken out of their hole in the bank of a river, and two of theſe conſtantly rejected flies, ants, earth worms, paſte and cheeſe, and died of hunger in two days; the two others ate a little cheeſe and ſome earth-worms, but lived only ſix days. Geſner obſerves that the King-fiſher can never be tamed, and is always equally wild. Its fleſh has the odour of baſtard muſk *, and is unpalatable food; its fat is reddiſh †; its ſtomach is roomy and flaccid, as in the birds of prey; and like them too it diſcharges by the bill the indigeſted fragments, ſcales and bones rolled into ſmall balls: the ſtomach is placed very low, and conſequently the æſophagus is very long ||; the tongue is ſhort, of a red or yellow colour, like the inſide and the bottom of the bill §.

It

* Tragus.

† Geſner.

|| Idem.

§ " On the ſeventh of July, 1771, ſays M. de Montbeillard, I received five young King-fiſhers (there had been ſeven of them in the neſt by the brink of a rivulet): they ate earth-worms, that were given to them. In theſe young King-fiſhers, the outer toe was ſo cloſely connected to the middle one as far as the laſt joint, as to give the appearance rather

It is ſingular, that a bird which flies ſo ſwiftly and with ſuch continuance, has not broad wings; they are, on the contrary, very ſmall in proportion to its bulk, and we may hence judge of the vaſt force of the muſcles that impel them. For no bird perhaps flies ſo rapidly; it ſhoots like an arrow: if it drops a fiſh from the branch on which it is perched, it often ſnatches it before it touches the ground. As it ſeldom ſits but on dry branches, it is ſaid to wither the tree on which it ſettles ‡.

It has been ſuppoſed, that the dried body of the Kingfiſher has the property of preſerving cloth and woollen ſtuffs from moths;

rather of a forked toe than of two diſtinct toes; the tarſus was very ſhort; the head was ſtriped acroſs with black and greeniſh blue; there were two fire-coloured ſpots, one on the eyes before, and the other longer under the eyes, which extending behind, becomes white; below the neck, near the back, the blue grows predominant, and a waving band of blue mixed with a little black runs the whole length of the body, and extends to the extremity of the coverts of the tail, where the blue becomes more vivid; the twelve quills of the tail were of an enbrowned blue; the twenty-two quills of the wings were each half brown, and half embrowned blue lengthwiſe; the breaſt was rufous ſhaded with brown; the belly whitiſh; the under ſide of the tail rufous inclined to orange; the bill was ſeventeen lines; the tail was very ſhort, broad, and pointed; the ventricle very capacious."

Obſervation communicated by M. de Montbeillard.

‡ Schwenckfeld.

and

and for this reaſon drapers hang it in their ſhops. Its ſmell of baſtard muſk may perhaps repel thoſe inſects, but the effect can be no greater than that of any other kind of penetrating odour. As the body dries eaſily, it has been alledged that the fleſh never putrefies *. But theſe imaginary virtues are nothing compared to the wonders related by ſome authors, who have collected the ſuperſtitious notions of the ancients on the ſubject of the Alcyon: it averted the thunder; it augmented hidden treaſure; and though dead it renewed its plumage each ſeaſon of moulting †: it beſtows, ſays Kirannides, on the perſon who carries it, gracefulneſs and beauty; it preſerves peace at home; it maintains calm at ſea; it draws together the fiſhes in abundance. ‡ Such fables charm the fondneſs for credulity;

* Geſner.

† Aldrovandus. *Lib. iii. p.* 621.

‡ What is ſingular, they are found alſo among the Tartars and Siberians. "The Kingfiſhers are ſeen over all Siberia, and the feathers of theſe birds are employed by the Tartars and the Oſtiacs, for many ſuperſtitious uſes. The former pluck them, caſt them into water, and carefully preſerve ſuch as float; and they pretend, that if with one of theſe feathers they touch a woman or even her cloaths ſhe will fall in love with them. The Oſtiacs take the ſkin, the bill, and the claws of this bird, and ſhut them in a purſe; and as long as they preſerve this ſort of amulet, they believe that

lity; but unfortunately they are only the offspring of an heated imagination. [A]

that they have no ill to fear. The perſon who taught me this mean of living happy, could not forbear ſhedding tears; he told me that the loſs of ſuch a ſkin that he had, made him loſe alſo his wife and his goods. I told him that ſuch a bird could not be ſo very rare, ſince a countryman of his had brought me one, with its ſkin and its feathers: he was much ſurprized, and ſaid, that if he had the luck to find one, he would give it to no perſon." *Voyage en Siberie*, par M. Gmelin, tom. ii. p. 112.

[A] Specific character of the King-fiſher, *Alcedo-Iſpida*: "It is ſhort-tailed, ſky-blue above, fulvous below, its ſtraps rufous." Mr. Pennant has given an excellent hiſtory of this bird, with critical diſcuſſions, in the Britiſh Zoology.

Foreign KING-FISHERS.

As the number of foreign ſpecies is very conſiderable, and as they are all inhabitants of the warm climates, we may regard the European King-fiſher, which occurs ſingle and detatched, as expatriated from the original ſtock. To give regularity to the enumeration of this multitude, we ſhall firſt ſeparate the King-fiſhers of the Ancient Continent from thoſe of America, and then range them in the order of their magnitude, beginning with ſuch as are larger than the European ſpecies.

GREAT KING-FISHERS of the Old Continent.

The GREATEST KING-FISHER.

First Species.

Alcedo Fusca. Gmel.
Alcedo Gigantea. Lath. Ind.
The Great Brown King-fisher. Lath. Syn.

THIS bird, which is the largest of its kind, occurs in New Guinea; it is sixteen inches long, and of the bulk of a Jackdaw; all the plumage, except the tail, appears stained with bistre*, and embrowned on the back and on the wing; the colour is lighter and faintly crossed by small blackish waves on all the foreside of the body and round the neck on a whiter ground; the feathers on the crown of the head, as well as a broad streak below the eye, are of the brown bistre of the back; the tail is of a rusty fulvous crossed with black waves, and is white at the end: the lower mandible is

* "A colour made of chimney soot boiled and then diluted with water; used by painters in washing their drawings."

orange,

orange, the upper one black, and ſlightly bent at the point; a character which, in ſome degree, detaches the bird from the King-fiſhers, though all its other properties agree with thoſe of that tribe.

The BLUE and RUFOUS KING-FISHER.

Second Species.

Alcedo Smyrnenſis Var. 1ſt. Linn. and Gmel.
Iſpida. Klein.
Iſpida Madagaſcarienſis Cærulea. Briſſ.
The Great Gambia King-fiſher. Edw. and Lath.

It is a little more than nine inches long, and its bill, which is red, meaſures two inches and a half; all the head, the neck, and the upperſide of the body, are of a fine brown rufous; the tail, the back, and half of the wings, are blue, varying according to its poſition into ſky-blue and ſea-green; the point of the wings and the ſhoulders are black. This ſpecies is found in Madagaſcar, it is alſo ſeen in Africa, on the river Gambia, according to Edwards. A King-fiſher from the Malabar coaſt, figured in the *Planches Enluminees*, and which

which Briſſon makes his fourteenth ſpecies *, reſembles exactly the preſent, except that its throat is white, a difference which may be imputed to diſtinction of ſex. In that caſe, it muſt inhabit the zone that ſtretches acroſs the whole continent; and if the Smyrna King-fiſher of Albin, which Briſſon reckons his thirteenth ſpecies †, be the ſame, as appears to us moſt probable, this bird is diſperſed through a ſtill greater extent. [A]

* *Alcedo Smyrnenſis Var. 2d.* Linn. and Gmel.
Iſpida Bengalenſis Major. Briſſ.
The Great Bengal Kingfiſher. Albin. and Lath.

† *Alcedo Smyrnenſis.* Linn. and Gmel.
Iſpida Smyrnenſis. Briſſ.

[A] Specific character: "It is long-tailed, ferruginous; its wings, its tail and its back green."

The CRAB KINGFISHER.

Third Species.

Alcedo Cancrophaga. Lath. Ind.
Alcedo Senegalenſis. Var. 1. Gmel.
The Crab-eating King-fiſher. Lath. Syn.

THIS King-fiſher was ſent us from Senegal under the name of *Crabier.* It is probably found alſo at the Cape de Verd iſlands,

iſlands, and the following indication, given by Forſter in Cook's ſecond voyage, ſeems to refer to it. "The moſt remarkable bird we ſaw at the Cape de Verd iſlands was a kind of King-fiſher, which lives upon the large red and blue land crabs that croud in the holes of the dry and parched ſoil*." The tail and all the back are of a ſea-blue, which alſo paints the outer margin of the great and middle quills of the wing, but their points are black, and a large ſpot of that colour covers all the part next the body, and marks on the wing the trace, as it were, of a ſecond wing; all the underſide of the body is light fulvous; a black ſtreak ſtretches behind the eye, the bill and legs are of a deep ruſt colour. The length of this bird is a foot.

* This obſerver adds: "The ſame ſpecies is found in Arabia Felix, and alſo in Abyſſinia, as appears from the elegant and precious drawings of Mr. Bruce."

The THICK-BILLED KING-FISHER.

Fourth Species.

Alcedo Capensis. Linn. and Gmel.
Ispida Capitis Bonæ Spei. Briss.
The Cape King-fisher. Lath.

THE bill of the King-fishers is in general large and strong: but in the present, it is uncommonly thick and stout. The whole length of the bird is fourteen inches; that of the bill alone is above three inches, and its thickness at the base eleven lines; the head is capped with light gray; the back is water-green; the wings are aqua-marine; the tail is of the same green with the back, and lined with gray; all the underside of the body is a dull weak fulvous; the bill has the red tint of Spanish wax. [A]

[A] Specific character of the *Alcedo Capensis:* "It is short-tailed, ash-blue, below fulvous, its breast brick colour, its bill red."

The PIED KING-FISHER.

Fifth Species.

Alcedo Rudis. Linn. and Gmel.
Ispida ex albo et nigro varia. Briss.
The Black and White King-fisher. Edw. and Lath.

THE whole plumage of this bird consists of black and white, broken and intermixed; and we have therefore termed it the *Pied King-fisher.* The back is of a black ground, clouded with white; there is a black zone on the breast; all the fore-side of the neck, as far as under the bill, is white; the quills of the wing are black on the outer edge, intersected within by white and black, and fringed with white: the top of the head and the crest are black; and the bill and legs of that same colour: The total length of the bird is near eight inches. [B]

This King-fisher came from the Cape of Good Hope. On comparing it with another sent from Senegal, and delineated in the *Planches Enluminees*, we cannot help regarding it one of the same species: since the

[B] Specific character of the *Alcedo Rudis:* "It is short-tailed, black variegated with whitish, below white."

the differences which occur in the two figures are inconſiderable; for inſtance, the black in the Senegal bird is lighter and more dilute; the feathers on the head which are repreſented lying flat are yet capable of riſing to a creſt: but the moſt material diſparity is, that it has a larger proportion of white, and the other from the Cape of Good Hope, has a ſomewhat larger proportion of black. Edwards gives one of theſe birds ſent from Perſia; but his figure is defective, and the colours altogether miſtaken. He tells us indeed that he received it prepared in ſpirits of wine, and obſerves himſelf, that after a long immerſion in that liquor the tints of the plumage are greatly weakened and diſordered. But it is quite improbable that the black and white variegated King-fiſher of Jamaica mentioned by Sloane, of which he gives a figure that merits hardly any attention, is the ſame ſpecies with that of Senegal or of the Cape of Good Hope; yet Briſſon makes no ſcruple in ranging them together. A bird which flies only ſhort diſtances, and ſkims along the ſhores, could never traverſe the vaſt Atlantic Ocean; and nature, ſo diverſified in all her operations, ſeems never to have repeated any of her forms in the other continent, but to have faſhioned her productions

tions after entirely new models. This is moſt likely an indigenous ſpecies, and appropriated entirely to thoſe countries where it occurs; and the ſame may be the caſe with the King-fiſhers ſeen in thoſe iſlands which are ſcattered in the midſt of the South Sea, and diſcovered by the lateſt navigators. Forſter, in his account of Captain Cook's ſecond voyage round the world, obſerves that they were found in Otaheite, Xuaheine, and Ulietea, which are more than one thouſand five hundred leagues diſtant from any continent. Theſe King-fiſhers are of a dull green, with a collar of the ſame about their white neck. It appears that ſome of the iſlanders entertain a ſuperſtitious veneration for theſe birds; and thus, from one end of the world to the other, have marvellous qualities been aſcribed to the family of the Alcyons *.

* In the afternoon we ſhot (at Ulietea) ſome King-fiſhers. The moment that I had fired laſt, we met with Oreo and his family, who were walking on the beach with Captain Cook. The Chief did not obſerve the bird which I held in my hand, but his daughter wept for the death of her *eatua* or genius, and fled from me when I offered to touch her: her mother and moſt of the women who accompanied her, ſeemed alſo concerned for this accident; and the Chief, mounting on his canoe, entreated us in a very ſerious tone to ſpare the King-fiſhers and the Herons of his iſland, at the ſame time granting us permiſſion to kill all the other birds. We tried in vain to diſcover the cauſe of this veneration for theſe two ſpecies." *Captain Cook's Second Voyage, by Forſter.*

The CRESTED KING-FISHER.

Sixth Species.

Alcedo Maxima. Var. Linn. and Gmel.
The Great King-fisher. Lath.

THIS King-fiſher is one of the largeſt, being ſixteen inches in length; its plumage is richly enamelled, though not marked with brilliant colours; it is entirely ſprinkled with white drops, ſtrewed in tranſverſe lines on a blackiſh gray ground, from the back to the tail; the throat is white with blackiſh ſtreaks on the ſides; the breaſt is enamelled with the ſame two colours, and with rufous; the belly is white; the flanks and the coverts under the tail are of a ruſt colour.

Sonnerat deſcribes a ſpecies of King-fiſher from New Guinea, which bears a great reſemblance to this, in its ſize and in part of its colours. We ſhall not however venture to decide on their identity.

The BLACK CAPPED KING-FISHER.

Seventh Species.

Alcedo Atricapilla. Gmel.

THIS is one of the moſt beautiful of the King fiſhers; a ſoft ſilky violet covers the back, the tail, and half the wings; their tips, and the ſhoulders are black; the belly is light rufous; a white plate marks the breaſt and the throat, and bends round the neck near the back; the head wears a large black cap; a great bill of a brilliant red completes the rich decoration of this bird: its length is ten inches. It is found in China; and we reckon the Great King-fiſher of the iſland of Luçon, mentioned by Sonnerat, as a contiguous ſpecies, or merely a variety.

The GREEN HEADED KING-FISHER.

Eighth Species.

Alcedo Chlorocephala.

A GREEN cap, with a black edging, covers the head; the back is of the ſame green, which melts on the wings and tail into a ſea-green; the neck, the throat, and all the fore ſide of the neck, are white; the bill, the legs, and the underſide of the tail are blackiſh: the length is nine inches. This bird, which appears to be of a new ſpecies, is repreſented in the *Planches Enluminées* as a native of the Cape of Good Hope; but we find from Commerſon's papers that he ſaw and deſcribed it in the iſland of Bouro, near Amboyna and one of the Moluccas.

The KING-FISHER, with Straw-coloured Head and Tail.

Ninth Species.

Alcedo Leucocephala. Gmel.
The White-headed King-fisher.

THIS is a new ſpecies: the wings and the tail are of a deep turquois-blue; the great quills of the wings are brown, fringed with blue; the back is of a ſea-green; the neck, the fore and under ſides of the body white, tinged with ſtraw or doe colour; ſmall black ſtrokes are traced on the white ground of the crown of the head; the bill is red and near three inches long; the total length of the bird is a foot. [A] The King-fiſher of Celebes, mentioned by voyagers, ſeems to belong to a ſimilar ſpecies, though rather ſmaller: but it is ſomewhat embelliſhed perhaps by their imagination. This bird, ſay they *, lives on a ſmall fiſh which it watches to catch on the river. It circles, razing the ſurface of the

[A] Specific character of the *Alcedo Leucocephala*: "It is blue-green; its head, its neck, and its under ſurface white; its wing-quills brown."

* Hiſtoire Generale des Voyages, *tom x. p.* 459.

water,

water, till its prey, which is very nimble, ſprings into the air, as if to dart down upon its enemy; but the bird is always dextrous enough to prevent the blow. It ſeizes the fiſh with its bill, and tranſports it to its neſt, where it ſubſiſts a day or two on its ſpoil, and ſpends the whole time in ſinging. It is ſcarcely bigger than a lark; its bill is red; the plumage on its head and back is entirely green; that of the belly verges on yellow, and its tail is of the fineſt blue in the world, This wonderful bird is called *Ten-rou joulon* *.

* This bird is reckoned by Gmelin and Latham a different ſpecies under the name of *Alcedo Flavicans*. Specific character: "Below yellowiſh, its head and back green; its bill and its tail blue."

The WHITE COLLARED KING-FISHER.

Tenth Species.

Alcedo Collaris.

SONNERAT deſcribes this new ſpecies. It is rather ſmaller than the Blackbird; its head, its neck, its wings, and its tail, are blue, ſhaded with green; all the underſide of the body is white, and there is a white a white ring round the neck. He found this bird in the Philippines, and we have reafon to believe that it occurs alfo in China.

The bird which Briffon defcribes, from a drawing, under the name of *the Collared Indian King-fifher*, and which he fays is much larger than our European King-fiſher, may be a variety of this tenth fpecies.

The Middle-ſized KING-FISHERS of the Ancient Continent.

The BABOUCARD.

Firſt Middle Species.

Iſpida Senegalenſis. Briſſ.
Alcedo Iſpida. Var. Gmel.

THE name of the King-fiſher at Senegal is in the language of the country *Baboucard.* The ſpecies are numerous on that great river *, and they are all painted with the moſt variegated and vivid colours. We apply the generic term Baboucard to Briſſon's ſeventh ſpecies, which reſembles ſo cloſely the European King-fiſher, that they may be regarded as contiguous kinds, or perhaps as really the ſame; ſince we have already obſerved that our own derives origin from the hot climates. [A]

* Adanſon.

[A] Specific character of the *Alcedo Senegalenſis:* "It is long-tailed, below ſky-blue, its head hoary, the coverts of its wings black."

The BLUE AND BLACK KING-FISHER of Senegal.

Second Middle Species.

Alcedo Senegalenſis. Var. 3. Gmel.

THIS appears to be rather larger than our King-fiſher, though it is ſcarcely ſeven inches long: the tail, the back, the middle quills of the wing, are deep blue; the reſt of the wing, including the coverts and the great quills, is black; the under-ſide of the body is ruſty fulvous as far as the throat, which is white, ſhaded with bluiſh; this tint a little deepened covers the upper ſide of the head and neck; the bill is rufous, and the legs reddiſh.

The GRAY HEADED KING-FISHER.

Third Middle Species.

Alcedo Senegalensis. Linn. and. Gmel.
The Senegal King-fisher. Lath.

THIS King-fisher is intermediate between the large and the small kinds. It is nearly of the size of the Throstle, being eight inches and a half long. Its head and neck are enveloped with brown gray, which is lighter and inclined to white on the throat and the foreside of the neck; the under surface of the body is white; all the upper surface sea-green, except a great black bar stretching on the coverts of the wing, and another which marks the great quills; the upper mandible is red, the lower black. [A]

[A] Specific character of the *Alcedo Senegalensis*: "It is long-tailed, sky-blue, below white, its head hoary, the coverts of its wings black.

The YELLOW FRONTED KING-FISHER.

Fourth Middle Species.

Alcedo-Erithaca. Linn. and Gmel.
Iſpida Bengalenſis Torquata. Briſſ.
The Bengal King-fiſher. Alb.
The Red-headed King-fiſher. Lath.

ALBIN gives an account of this bird, and tells us that it is the ſize of the Engliſh King-fiſher. If we may truſt more to the deſcriptions of this author than to his engravings, this ſpecies is diſtinguiſhed from the reſt by the beautiful yellow which tinges all the underſide of the body and the front; a black ſpot riſes at the bill and ſurrounds the eyes; behind the head there is a bar of dull blue, and then a ſtreak of white; the throat alſo is white; the back deep blue; the rump and tail dirty red; the wings, faint iron gray. [A]

[A] Specific character of the *Alcedo Erithaca*: "It is ſhort-tailed, its back blue, its belly yellow, its head and rump purple, its throat and nape white."

The LONG SHAFTED KING-FISHER.

Fifth Middle Species.

Alcedo-Dea. Linn. and Gmel.
Ispida Ternatana. Briss.
Pica Ternatana. Klein.
The Ternate King-fisher. Lath.

THIS ſpecies is diſtinguiſhed by a remarkable character; the two middle quills of the tail project and taper into two long ſhafts, which are naked three inches of their length, and have a ſmall beard of feathers at their ends; ſoft and deep turquois blue, and black velvet brown cover and interſect the upper ſurface with four large ſpots; the black occupies the top of the back and the point of the wings, the deep blue their middle, the upper part of the neck and head; all the under ſurface of the body and tail is white, lightly tinged with dilute red; the bill and legs are orange, on each of the two feathers in the middle of the tail, there is a blue ſpot, and the long ſhafts are of the ſame colour. Seba calls this bird, on account of its beauty, *The Nymph of Ternate*: he adds, that the feathers of the tail are one third longer in the male than in the female.

[A] Specific character of the *Alcedo-Dea*: "Two of its tail quills are very long, attenuated in the middle, its body dark-bluiſh, its wings greeniſh."

Small KING-FISHERS of the Ancient Continent.

The BLUE HEADED KING-FISHER.

First Small Species.

Alcedo Cæruleocephala. Gmel.

SOME King fishers are as small as the Gold Crested Wren, or, to compare them with a family that bears more affinity to them, they are as small as the Todies. That from Senegal is of this number; it is scarcely four inches long; all the under surface of the body, as far as the eye, is of a fine rufous; the throat is white; the back is of a fine ultramarine blue, the wing is of the same blue except the great quills which are blackish; the crown of the head is of a bright blue, stained with small waves of a lighter and verdant blue; the bill is very long in Proportion to the body, being thirteen lines. This bird was sent to us from Madagascar.

The RUFOUS KING-FISHER.

Second Small Species.

Alcedo Madagascariensis. Linn. and Gmel.
Ispida Madagascariensis. Briss.

IT is scarce five inches long; all the upper surface of the body from the bill to the tail, is of a bright shining rufous, except that the great quills of the wing are black, and the middle ones only fringed with that rufous on a blackish ground; all the under surface of the body is white tinged with rufous; the bill and legs are black. Commerson saw and described it at Madagascar.

The PURPLE KING-FISHER.

Third Small Species.

Alcedo Purpurea. Gmel.

IT is of the same size with the preceding: of all these birds it is the handsomest, and perhaps the richest in colours; a fine aurora rufous, clouded with purple, intermingled with blue, covers the head, the rump,

rump, and the tail; all the under ſide of the body is gold rufous, on a white ground; its mantle is enriched with an azure blue on a velvet black; a ſpot of light purple riſes at the corner of the eye, and terminates behind in a ſtreak of the moſt vivid blue; the throat is white, and the bill red. This charming little bird was brought from Pondicherry.

The WHITE BILLED KING-FISHER.

Fourth Small Species.

Alcedo Leucoryncha. Gmel,
Iſpida Americana Cœrulea. Briſſ.
Iſpida roſtro albo. Klein.
Alcedo Americana, ſeu Apiaſtra. Seba.

SEBA, from whom we borrow the account of this little King-fiſher, ſays that its bill is white, its neck and head of a red bay tinged with purple; the ſides are coloured the ſame; the quills of the wing are cinereous; their coverts and the feathers of the back are of a very fine blue, the breaſt and belly are light yellow: its length is about four inches and a half. When Seba

 aſſerts

aſſerts that theſe birds of the Alcyon tribe live on bees, he confounds them with the Bee-eaters; and, for the ſame reaſon, Klein corrects a capital error of Linnæus, who takes the *Iſpida* for the *Merops*; whereas the latter inhabits the wild tracts near forreſts, and not the margin of ſtreams, where it would find no bees. But Klein himſelf is guilty of an inaccuracy, in ſaying that the Alcyon of Seba appears like our King-fiſher; ſince, beſides the diſparity of bulk, the colours of the head and bill are totally different.

Voſmaer has deſcribed two ſmall King-fiſhers * which he refers to this Alcyon of Seba, but at the ſame time aſſerting that they have *only three toes*, two before and one behind. This fact required to be proved, and has been ſo by an excellent obſerver, as we ſhall afterwards find.

* *Alcedo Tridactyla.* Linn. and Gmel.

The BENGAL KING-FISHER.

Fifth Small Species.

Alcedo Bengalenſis. Gmel.
Iſpida Bengalenſis. Briſſ.
The Little Indian King-fiſher. Edw. and Lath.

EDWARDS gives in the ſame plate two ſmall King-fiſhers which appear to be cloſely related ſpecies, or perhaps the male and female of the ſame, though Briſſon makes them two ſeparate ſpecies. They are not larger than the Todies; in the one the upper ſurface is ſky-blue, and in the other ſea-green; the quills of the wings and of the tail are brown gray, in the former, and of the ſame green with the back, in the latter; the under ſide of the body is orange fulvous in both. Klein ſays that this ſpecies is like the European in its colours; he might have obſerved that it differs widely in ſize: but always impreſſed with the falſe notion that the toes are placed *two and two* in the genus of the King-fiſhers, he complains that Edwards is not ſufficiently diſtinct on that point, though the figures of this naturaliſt are here, as uſual, very accurate and well delineated.

The THREE TOED KING-FISHER.

Sixth Small Species.

Alcedo Tridactyla. Var. Linn. and Gmel.

We have already found in the Woodpeckers a ſingularity of this nature in regard to the number of the toes: this is leſs ſurprizing in the family of the King-fiſhers, where the little inner toe is ſo ſhort and almoſt uſeleſs, that it may be eaſily omitted. We are indebted to Sonnerat for the account of this ſmall three-toed King-fiſher, which is one of the richeſt and moſt brilliant in regard to plumage of the genus: all the upperſide of the head and back is of a deep lilac; the feathers of the wings are of a dull indigo, but heightened by a border of vivid and ſhining blue, which ſurrounds each feather; all the underſide of the body is white; the bill and legs are reddiſh. Sonnerat found this bird in the iſland of Luçon. Voſmaer ſays merely that his ſpecimens came from the Eaſt Indies.

We regard this ſpecies, the preceding of Seba and that of our purple King-fiſher, as as three contiguous ſpecies, and which might be reduced to two, or even to one, if it were eaſier to eſtimate the arbitrary differences of deſcriptions, or to rectify them from the objects themſelves. Voſmaer gives, under the name of Alcyon, two other birds which are not King-fiſhers; the firſt, which he calls the *Long-tailed American Alcyon*, beſides that its tail is too long in proportion, is excluded from the genus by the curvature of its bill; the ſecond has a longiſh, ſlender, quadrangular bill, and its toes bent two and two, and is therefore a Jacamar †.

† In a long note, which we ſhall omit, our author expoſes the peculiarities, the inelegancies, and the miſtakes of Voſmaér.

The VINTSI.

Seventh Small Species.

Alcedo Criſtata. Linn. and Gmel.
Iſpida Philippenſis Criſtata. Briſſ.
Iſpida roſtro luteo. Klein.
The Creſted King-fiſher. Lath.

VINTSI is the name given by the inhabitants of the Philippines to this ſmall King-fiſher, which thoſe of Amboyna, according to Seba, term the *Toborkey* and *Hito*,

Hito. The upper ſurface of the wings and the tail are ſky blue; the head is thick covered with long narrow feathers, neatly dotted with black and greeniſh points, that riſe to a creſt; the throat is white; on the ſide of the neck, there is a tawny rufous ſpot; all the under ſurface of the body is dyed with this colour: the whole bird is hardly five inches long.

The ſeventeenth ſpecies of Briſſon appears to be much related to the preſent, if it is not entirely the ſame; for the ſlight difference that occurs ſeems to indicate no more at leaſt than a variety. We cannot aſcertain to what ſpecies the ſmall bird of the Philippines, which Camel calls *Salaczac**, and which appears to be a King-fiſher; but nothing more than the name is given in the enumeration of the Philippine birds inſerted in the Philoſophical Tranſactions.

Briſſon deſcribes alſo another ſpecies of the King-fiſher, from a drawing that was ſent him from the Eaſt Indies; but as we have not ſeen the bird, we can add nothing to his delineation.

* Latham ſuppoſes it to be a variety of the *Vintſi*: it is the *Alcedo criſtata elegantiſſime picta* of Seba, and the *Iſpida Indica Criſtata* of Briſſon, mentioned in the next paragraph.

The KING-FISHERS of the New Continent.

Great Species.

The TAPARARA.

First Great Species

Alcedo Cayanensis. Gmel.
Ispida Cayanensis. Briss.
The Cayenne King-fisher. Lath.

TAPARARA is the generic name of the King-fisher in the language of the natives of Cayenne; and we shall apply it to this species, which is found in that island. It is as large as a Stare; the upper side of the head, the back, and the shoulders, are of a fine blue; the rump is of a sea-green; all the underside of the body is white; the quills of the wing are blue without, black within and below; those of the tail are the same, except that the two mid-ones are entirely blue; below the back of the head, there is a black cross bar. The vast abundance of water

water in the country of Guiana is favourable to the multiplication of the King-fiſhers. Accordingly, their ſpecies are numerous, and ſerve to point out the rivers that are ſtored with fiſh, they being frequently found by their banks. There are many King-fiſhers, ſays M. de la Borde, on the river Ouaſſa ; but they never congregate, and always appear ſingle; they breed in thoſe regions, as they do in Europe, in the perpendicular banks ; there are always a number of theſe holes near each other, though each of their lodgers lives ſolitary. M. de la Borde ſaw their young in the month of September, which in that climate is perhaps the time of incubation ; the cry of theſe birds is *carac, carac*. [A]

[A] Specific character of the *Alcedo Cayanenſis* : " It is blue, below white, a tranſverſe black bar below the back of the head."

The ALATLI.

Second Great Species.

Alcedo Torquata. Linn. and Gmel.
Iſpida Mexicana Criſtata. Briſſ.
The Cinereous King-fiſher. Lath.

WE form this name by contraction for *achalalactli* or *michalalactli*, which the bird receives in Mexico, according to Fernandez. It is one of the largeſt of the King-fiſhers, being near ſixteen inches long; but its colours are not ſo brilliant as thoſe of the others: bluiſh gray is ſpread over all the upperſide of the body, and that colour is variegated on the wings with white fringes in feſtoons at the point of the quills, the largeſt of which are blackiſh, and interſected within by broad white indentings; thoſe of the tail are marked with broad ſtripes of white; the underſide of the body is cheſnut-rufous, which grows more dilute as it riſes on the breaſt, where it is ſcaled or mailed with gray; the throat is white, and that colour, extending on the ſide of the neck, makes an entire circuit, and hence Nieremberg calls it *the collared bird*; all the head and the nape of the neck

neck are of the ſame bluiſh gray with the back. This is a migratory bird; it arrives at a certain time of the year in the northern provinces of Mexico, and probably comes from the hotter regions, for it is found in the Antilles, and we received it from Martinico. Adanſon ſays, that *it occurs likewiſe, though rarely, in Senegal, in the places near the mouth of the Niger.* But the difficulty of ſuppoſing that a Mexican bird could be found alſo in Senegal ſtruck himſelf, and he ſought to trace the differences between the *Achalalactli* of Fernandez and Nieremberg and this African Kingfiſher; and it thence appears that the bird deſcribed by Briſſon, and delineated in our *Planches Enluminees* is not the real Mexican Achalalactli, but that of Senegal. Indeed the diſtance between the climates is ſo vaſt, that we cannot doubt that birds unable to perform a long paſſage muſt be different ſpecies. [A]

[A] Specific character of the *Alcedo Torquata*: "It is ſhort-tailed, half-creſted, hoary bluiſh, with a white collar, its wings and tail ſpotted with white."

The JAGUACATI.

Third Great Species.

Alcedo-Alcyon. Var. 3d.
Ispida Brasiliensis Cristata. Briss.

WE have seen that the European species of King fisher occurs in Asia, and occupies perhaps the whole extent of the ancient Continent: the present is another which is found from the one extremity to the other in the new, from Hudson's Bay to Brazil. Marcgrave has described it under the Brasilian name *Jaguacati-guacu*, and the Portuguese appellation *Papapeixe*. Catesby saw it in Carolina, where he says it preys both on lizards and on fish. Edwards received it from Hudson's Bay, where it appears in the spring and summer; Brisson introduces it three times from these three authors, without comparing them, though the resemblance is obvious and remarked by Edwards himself. We have received this bird from St. Domingo and from Louisiana: some slight differences may be perceived; the most material one is that the scarf of the throat is marhed with little infous festoons in that from St. Domingo,

mingo, but is merely gray in the other, and the tail of the former appears ſomewhat more dotted and regularly ſprinkled with drops on all the quills, which drops are leſs viſible in the latter, and never appear except when the tail is ſpread: the compaſs of the neck is white, and alſo the throat; there is ſome rufous on the breaſt and on the ſides; the quills of the wing are black, marked with white at the point, and interſected in the middle by a ſmall white fringe, and which is only the border of the indentings on the inner webs, which appear when the wing is expanded. Marcgrave compares the bulk to that of the Fieldfare. Klein, who was not acquainted with the Great King-fiſhers of New Guinea, takes this for the largeſt of the kind *.

* It neſtles in high banks, into which it penetrates deep in a horizontal direction. It lays four eggs, and hatches in June. It ſeems to migrate from the northern parts of America to Mexico; where it is eaten, though it has a rank fiſhy taſte. T.

The MATUITUI.

Fourth Species.

Alcedo Maculata. Linn. and Gmel.
Ispida Brasiliensis Nævia. Briss.
The Brasilian spotted King fisher. Lath.

MARCGRAVE also describes this Brasilian King-fisher, and marks its true characters: the neck and legs, short; the bill, straight and strong, its upper mandible vermillion, extending over the lower, and bending somewhat at its point, a peculiarity observed already in the King-fisher of New Guinea. It is as large as the Stare; all the feathers of the head, of the upper-side of the neck, of the back, and of the tail, are fulvous or brown, spotted with yellowish white, as in the Sparrow-hawk; the throat is yellow, the breast and belly are white, dotted with brown. Marcgrave mentions nothing particular of its natural habits. [A]

In Fernandez and Nieremberg we find some birds which are improperly termed

[A] Specific character of the *Alcedo Maculata*: "It is brown, spotted with yellowish; below white, spotted with brown; its throat, bright yellow."

King-

King-fiſhers. Such are 1. the *hoactli* whoſe legs are a foot long, and therefore is by no means a King-Fiſher: 2. the *axoquen* whoſe neck and legs are equally long: 3. the *acacahoactli*, or the *aquatic bird with a raucous voice* of Nieremberg, *which ſtretches and bends back its long neck*, and appears to be a kind of Stork or Jabiru, much like the *hoacton*, which Briſſon terms *the creſted Mexican heron*. We may ſay the ſame of the *tolcomoctli* and *hoexocanauhtli* of Fernandez, which reſemble more this genus, but have ſome habits oppoſite to thoſe of the King-fiſhers *, though the Spaniards name them, as they do the preceding birds, *Martinetes Peſcadors* (Martin-fiſhers): Fernandez obſerves that the ſame appellation has been beſtowed on theſe very different ſpecies, merely becauſe they all live on fiſh.

* Fernandez ſays of the firſt, *that the ſtroke of its bill is dangerous*; this cannot be a King-fiſher, which is an innocent timid bird: and of the ſecond, *that it neſtles among the willows*; but all the King-fiſhers which have been obſerved, breed in the banks of ſtreams.

Middle-ſized KING-FISHERS, of the New Continent.

The GREEN and RUFOUS KING-FISHER.

Firſt Middle Species.

Alcedo Bicolor. Gmel.

THIS King-fiſher is found in Cayenne: all the underſide of the body is of a deep gold rufous, except a zone waved with white and black on the breaſt, which diſtinguiſhes the male; a ſmall ſtreak of rufous extends from the noſtrils to the eyes; all the upperſide of the body is of a dull green, ſprinkled with ſome ſmall whitiſh ſpots, thinly ſcattered; the bill is black and about two inches long; the tail meaſures two inches and a half, which gives the bird a length of eight inches, though it is not thicker than the common King-fiſher.

The GREEN and WHITE KING-FISHER.

Second Middle Species.

Alcedo Americana. Gmel.

THIS ſpecies too occurs in Cayenne: it is ſmaller than the preceding, being only ſeven inches long, though its tail is ſtill of conſiderable length; all the upper-ſide of its body is gloſſed with green on a blackiſh ground, interſected only by a white horſe-ſhoe, which, riſing under the eye, deſcends on the back of the neck, and by ſome white ſtreaks thrown on the wing; the belly and ſtomach are white, and variegated with ſome ſpots of the colour of the back; the breaſt and the fore part of the neck are of a fine rufous in the male, and this character diſtinguiſhes it from the female, which has a white throat.

The GIP-GIP.

Third Middle Species.

Alcedo Brasiliensis. Gmel.
Ispida Brasiliensis. Briss.

THIS is the *anonymous* bird of Marcgrave, which may be called *Gip-gip* on account of its cry. It is as large as the Lark, and of the form of the *Matuitui*, which is the fourth great species of American King-fishers; its bill is straight and black; all the upperside of the head, neck, wings and tail, is reddish, or rather shady-bay, mixed with white; the throat and the underside of the body, are white, and a brown streak runs from the bill to the eye: its cry *gip gip* resembles the puling of young turkeys.

SMALL KING-FISHERS of the New Continent.

The GREEN and ORANGE KING-FISHER.

Alcedo Superciliosa var. Linn. and Gmel.
Ispida Americana Viridis. Briss.

THIS is the only species in America which may be termed a *Small King-fisher*, and it is scarce five inches long: all the underside of the body is of a brilliant orange, except a white spot on the throat, another on the stomach, and a deep green zone below the neck in the male, and which is wanting in the female; both of them have a half-collar of orange behind the neck; the head and all the upper surface, are covered with green-gray, and the wings are spotted with small rusty drops near the shoulder and on the great quills which are brown. Edwards, who gives the figure of this bird *, says that he could not discover from what country it was brought; but we received it from Cayenne.

* The little green and orange-coloured King-fisher. *Gleanings*, pl. 245.

The JACAMARS.

We have formed this name by ſhortening the Braſilian appellation *Jacamaciri*. Theſe birds differ not from the King-fiſhers, except that their toes are diſpoſed two before and two behind, while thoſe of the King-fiſhers are placed three before and one behind. But the Jacamars reſemble them in the ſhape of their body and of their bill, and they are of the ſame ſize with the middle ſpecies of King-fiſhers; and this is probably the reaſon that ſome authors * have ranged them together. Others † have claſſed the Jacamars with the Woodpeckers, the diſpoſition of their toes being ſimilar, and the ſhape of their bill nearly the ſame, though longer and more ſlender; but they are diſcriminated from the Woodpeckers, ſince their tongue is not longer than their bill, and the feathers of their tail are neither ſtiff, nor wedge-ſhaped. It appears therefore, that the Jacamars conſtitute a ſeparate genus, which has as great affinity to

* Edwards, &c.

† Willughby, Klein, &c.

the Woodpeckers, perhaps, as to the King-fishers; it contains only two species, which are both natives of the hot climates of America.

The JACAMAR, properly so called.

First Species.

Alcedo Galbula. Linn. and Gmel.
Galbula. Briss.
Galbula Viridis. Lath. Ind.
The Cupreous Jacamar. Penn.
* *The Green Jacamar.* Lath. Syn.

THIS bird is about the size of a lark, and its whole length is six and a half inches; the bill is an inch and five lines; its tail, only two inches, yet it projects an inch beyond the wings, when they are closed; the quills of the tail are very regularly tapered; the legs are very short, and of a yellowish colour; the bill is black, and the eyes are of a fine deep blue; the throat is white, and the belly rufous; all the rest of the plumage is of a very brilliant gold-green, with red copper reflections.

* The savages of Cayenne call this bird *Venetou*; and the Creoles give it the appellation *Colibri des grands bois* (the Forest-Colibri).

In

No. 171

THE GREEN JACAMAR.

In ſome ſubjects, the throat is rufous, as well as the belly; in others, the throat is only a little yellowiſh; the colour of the upperſide of the body, alſo, is more or leſs brilliant in different ſpecimens, which may be attributed to age or ſex.

Theſe birds are found both in Guiana and in Brazil; they inhabit the foreſts, and prefer the wet places, as affording in moſt abundance their inſect food; they never join in ſociety, but conſtantly reſide in the moſt ſequeſtered and darkeſt coverts; their flight, though rapid, is ſhort; they perch on the middle boughs, and remain at reſt the whole of the night and the greateſt part of the day; they always are alone, and almoſt perpetually tranquil: yet there is uſually a number in the ſame diſtrict, that make reſponſes in a feeble broken warble, but which is tolerably pleaſant. Piſo ſays that their fleſh, though hard, is eaten in Braſil. [A]

[A] Specific character of the *Alcedo-Galbula*: "Its tail is wedge-ſhaped, its body green gold, below rufous, its feet ſcanſory."

The Long-tailed JACAMAR.

Second Species.

Alcedo Paradisea. Linn and Gmel.
Galbula Longicauda. Briss.
Galbula Paradisea. Lath. Ind.
Ispida Surinamensis. Klein.
The Swallow-tailed King-fisher. Edw.
The Paradise Jacamar. Lath. Syn.

THIS bird is rather larger than the preceding, from which it differs by its tail having twelve quills, while in that of the other there are only ten; the two middle quills of the tail are besides much longer, exceeding the rest two inches and three lines, and measuring in all six inches. It resembles the former Jacamar however in the form of its body and of its bill, and in the disposition of its toes; yet Edwards gives it three toes before and one behind, and this mistake probably has induced him to reckon it a King fisher. It differs from the first Jacamar by the tints and distribution of its colours, which have nothing common to both but the white on the breast; all the rest of the plumage is of a dull and deep green, in which we distinguish only some orange and violet reflections.

We

We are unacquainted with the female of the preceding ſpecies; but that of the preſent is diſtinguiſhed from the male by the two middle quills of the tail, which are much ſhorter, nor has its plumage any of the orange and violet reflections.

Theſe long-tailed Jacamars live on inſects like the others. But all their other habits perhaps differ; for they ſometimes frequent the cleared grounds, they fly to great diſtances, and they perch on the tops of trees; they go alſo in pairs, nor are they ſo ſolitary or ſedentary as the others; they have not the ſame warble, but a cry or rather a ſoft whiſtle, which is heard only when near, and is ſeldom repeated. [A]

[A] Specific character of the *Alcedo Paradiſea*: "Its two middle tail-quills very long, its body green gold, its feet ſcanſory."

The TODIES.

Les Todiers. Buff.

SLOANE and Browne are the firſt who have deſcribed one of theſe birds, which they term *Todus.* But beſides this ſpecies from Jamaica, we know two or three others, which all ſeem to be natives of the hot climates of America. The diſcriminating character of the genus is, that as in the King-fiſhers and the Manakins, the mid-toe is cloſely connected, and as it were glued to the outer-toe as far as the third joint, but cohering to the inner-toe in the ſame way only at the firſt joint. If we reſted on this property therefore, we ſhould claſs the Todies with the Manakins or King-fiſhers; but they are diſtinguiſhed from theſe and indeed from all other birds, by the form of the bill, which is long, ſtraight, blunt at the end, and flattened above and below, ſo that they have been called by the Creoles of Guiana, *Little Pallets* or *Little Spatulas.* This ſingular confirmation of their bill is alone ſufficient to conſtitute a diſtinct genus.

Nº 172

THE GREEN TODY.

The NORTH AMERICAN TODY.

First Species.

Todus Viridis. Linn. and Gmel.
Sylvia Gulâ Phœniceâ. Klein.
Todus Viridis pectore rubro. Browne.
Rubecula Viridis elegantissima. Ray and Sloane.
The Green Sparrow, or Green Humming-bird. Edw.
The Green Tody. Penn. and Lath.

THIS Tody is not larger than the Gold-Crested Wren, being about four inches long. We shall not here copy the long descriptions given by Browne, Sloane, and Brisson; because it will always be easy to distinguish the bird; for, besides the peculiarity of the bill, the upper side of the body, in the male, is of a dilute blue, and the under side rose colour; and, in the female, the back is of a fine green, and the rest of the plumage similar to that of the male. In both, the bill is reddish, but lighter below and browner above; the legs are gray, and the nails long and hooked. The bird feeds on insects and small worms, and innabits wet and sequestered spots. The two subjects designed in the *Planches Enlumi-*

nees

nees were ſent us from St. Domingo by M. Chervain, under the name of *Land Parrots*, but with the deſcription of the female only. He obſerves that, in the love ſeaſon, the male has a feeble, though pleaſant, warble; that the female builds her neſt on the dry ground, and preferably on the friable mold; and, for that reaſon, theſe birds chuſe the ravines and water-gullies; they often neſtle alſo in the low galleries of houſes, yet always on the ground; they make an excavation with their bill and claws, give it a round form, hollow out the bottom, and place pliant ſtraws, dry moſs, cotton, and feathers, which they artfully arrange; they lay four or five eggs, which are gray and ſpotted with deep yellow.

They catch, with great dexterity, flies and other ſmall winged inſects; they are difficult to tame, yet we may ſucceed if they are young, and fed by their parents in a cage until they can eat by themſelves; they have a ſtrong attachment to their brood, and will not leave them as long as they hear them cry.

We have ſeen that Sloane and Browne found this bird in Jamaica; it occurs alſo in Martinico, whence M. de Chanvalon ſent it to M. de Reaumur. It appears therefore that

that this ſpecies inhabits the iſlands and continent in the warmeſt parts of North America; but we cannot diſcover whether it alſo reſides in South America, at leaſt Marcgrave makes no mention of it.

[A] Specific character of the *Todus Viridis*: "It is green, its breaſt red."

The TIC-TIC, OR SOUTH AMERICAN TODY.

Second Species.

Todus Cinereus. Linn. and Gmel.
The Gray and Yellow Fly-catcher. Edw.
The Cinereous Tody. Lath.

THE natives of Guiana call this bird *Tic-tic*, in imitation of its cry; it is as ſmall as the preceding, which it reſembles exactly in the ſhape of its bill, and in the conformation of its toes; it differs in its colours, being cinereous mixed with deep blue on the upper ſide of the body, whereas the former is of a light ſky-blue on the ſame parts: this difference in the ſhade of the colours would only mark a variety; but all the under ſide of the body is yellow, there is no roſe-colour on the throat or on the flanks, and as the bird belongs to a different climate, we reckon it a diſtinct ſpecies.

ſpecies. It differs from the North American Tody, alſo, becauſe the tips of the lateral quills of the tail are white, for the ſpace of five or ſix lines; yet this property is peculiar to the male, for in the female, the extreme quills are of an uniform colour, and of the ſame aſh gray with the upper ſide of the body; all theſe tints are alſo fainter and more dilute than in the male.

This bird lives on inſects, like the preceding; it prefers the cleared grounds for its haunts; it is ſcarcely ever found in the foreſts, but often among thickets and buſhes.

[A] Specific character of the *Todus Cinereus*: "It is cinerous, below yellow."

The ORANGE BELLIED BLUE TODY.

Third Species.

Todus Cœruleus. Gmel.
The Blue Tody. Lath.

WE have cauſed this Tody to be delineated from a ſpecimen in good preſervation in the cabinet of M. Aubry, rector of St. Louis: it is three inches and ſix lines in length; the upper part of the head, neck,

neck, and all the back, are of fine deep blue; the tail and the tips of the wing-coverts are of the ſame colour; all the under ſurface of the body; and alſo the ſides of the head and neck, are of a fine orange, the lower part of the throat is whitiſh; near the eyes are ſmall daſhes of violet purple. This deſcription will ſuffice to diſtinguiſh this Tody from others of the kind.

There is a fourth bird which Briſſon has deſcribed from Aldrovandus under the name of the *Variegated Tody* *, and we ſhall here condenſe the account given by theſe two authors. It is of the ſize of the Gold Creſted Wren; its head, throat and neck are blackiſh blue, the wings green, the quills of the tail black edged with green, and the reſt of the plumage variegated with blue, black and green. But as Briſſon takes no notice of the ſhape of the bill, and as Aldrovandus, who is the only perſon that has ſeen this bird, is equally ſilent on that point, I cannot decide whether it really belongs to the genus of the Todies. [A]

* *Todus Varius.* Gmel.
Iſpida Indica. Aldrov.
The Variegated Tody. Lath.

[A] See note at the end of No. 8.

The AQUATIC BIRDS.

THE Aquatic are the only claſs of birds, which, to their common inheritance of the air and of the earth, join alſo the poſſeſſion of the ſea. Numerous ſpecies, containing vaſt multitudes of individuals, inhabit its ſhores and its level ſurface; they float on the billows with as much eaſe and with more ſecurity than they ſoar in their native element. Their proviſions are ever abundant, their prey cannot eſcape their purſuit: ſome plunge into the waves, others only ſweep the face of the water: all of them dwell on the fluctuating face of the deep, as if it were a ſtable abode; they form a great ſociety, and live in tranquillity amidſt the ſtorms: they ſeem even to play with the billows, to contend with the winds, and to expoſe themſelves to the vehemence of the tempeſt, without apprehending or ſuffering ſhipwreck.

It is with reluctance that they leave their favourite reſidence, and never until the concerns of incubation detain them on ſhore, or permit only ſhort excurſions into the ſea; but, as ſoon as their young are hatched,

hatched, they introduce them into their proper element: there, they may continue as long as they chuse; no water can penetrate their plumage, and when fatigued by flying, they may recruit their exhausted vigor by resting on the surface. The long dark nights, or the continued violence of storms * are the only hardships to which they are exposed, and which at times oblige them to retire to the shore. Then they announce to the navigator the proximity of the land, and serve, by their flight, to guide his course. Yet Captain Cook advises to regard the appearance of these birds as no certain sign †, since they often rove at vast distances on the main; and it appears from the accounts of mariners that the greater

* "The disorder of the elements (in a great storm) drove not the birds from us: from time to time a *Black Tern* fluttered on the troubled face of the sea, and broke the force of the waves by exposing itself to their action. The aspect of the ocean was then threatning and terrible." *Forster.*

† "The Blue Petrels which are found on this immense sea are no less insensible to cold than the Penguins. We have seen them between New Zealand and America, more than seven hundred leagues from any land." *Forster.* "We frequently had occasion to remark that these birds give not more certain signs of land than the sea-weed, except those species which never rove very far from the coasts. With regard to the Penguins, Petrels, and Albatrosses, as we meet with them six or seven hundred leagues at sea, we cannot reckon at all upon their indication." *Id.*

 number

number do not return each night to the beach, but repoſe among the ſhelves, or ſlumber on the water ‡

The ſhape and confirmation of theſe birds ſhew ſufficiently that they are deſtined by nature to inhabit the watry element; their body is arched and bulged like the hulk of a ſhip, and this figure was perhaps copied in the firſt conſtruction of veſſels; their neck, which riſes on a projecting breaſt, repreſents the prow; their ſhort tail, collected into a ſingle bundle, ſerves as a rudder §; their broad and palmated feet perform the office of oars; and their thick down, gliſtening with oil, which entirely inveſts them, is impenetrable by humidity, and, at the ſame time, enables them to float more lightly on the ſurface of the water ||. The habits and œconomy of theſe birds correſpond alſo to their organization; they never ſeem happy but in their appropriated element; they are averſe

‡ "There is reaſon to ſuppoſe that they even ſleep upon the water. We paſſed near an Albatroſs that was reſting aſleep on the ſurface, having been fatigued by the preceding ſtorm." *Forſter.*

§ Ariſtotle, *Hiſt. Anim.* Lib. ii. 5.

|| "The birds of warm countries are moderately cloathed, while thoſe of cold countries, particularly ſuch as fly inceſſantly on the ſea, have a prodigious quantity of feathers, each of which is double."

averſe to alight on the land; and the leaſt roughneſs of the ground hurts their ſoles, which are ſoftened by the perpetual bathing. The water is to them the ſcene of pleaſure and repoſe, where all their motions are performed with facility, and their various evolutions traced with elegance and grace. View the Swans moving ſweetly along, or ſailing majeſtically with expanded plumage upon the wave; they gaily ſport, they dive and again emerge with gentle undulations, and ſoft energy expreſſive of thoſe ſentiments which are the foundation of love: the Swan is the emblem of gracefulneſs, the quality which firſt commands our attention, even prior to the faſcination of beauty.

The life of the Aquatic Birds therefore is more peaceful and leſs laborious than that of moſt other tribes; ſmaller force is required in ſwimming than is expended in

"It is a miſtake to aſcribe to the *Alcyon* alone the inſtinct of following veſſels: ſince many ſea-birds paſs the greateſt part of their life on this element, far from the ſhores, and they can ſcarce poſſibly find ſubſiſtence during a ſtorm on the troubled face of the deep: they then keep in the veſſels wake, often before the wind comes on, and feed upon what is thrown over-board. Beſides they can repoſe on the ſmooth track which the veſſel leaves." *Remarks made by the Viſcount de Querhoent.*

N. B. This *Alcyon* of the mariners is different from the true Alcyon of the ancients or the King-fiſher; it is probably a ſea-ſwallow.

 flying;

flying; and the element which they inhabit perpetually yields them ſubſiſtence: they rather light on their prey than ſearch for it, and often a friendly wave conveys it within their reach, and they ſeize it without trouble or fatigue. Their diſpoſitions alſo are more innocent, and their habits more pacific. Each ſpecies congregates, from mutual attachment; they never attack their companions nor deſtroy other birds; and, in this great and peaceful nation, the ſtrong never oppreſs the weak. Very different from thoſe tyrants of the air and of the land, which ſpread through their empire diſorder, devaſtation, and war; the winged inhabitants of the water, live in profound harmony with each other, and are never polluted by the blood of their kindred: they reſpect even the whole claſs of birds, and employ their ſtrength and their weapons only againſt the abject ſwarms of inſects, and the dumb tribes of fiſhes. Yet moſt of the Aquatic Birds have a keen appetite, and are furniſhed with arms correſponding. Many ſpecies, ſuch as the Gooſander, the Brent-gooſe, the Shell-drake, &c. have the inner edges of their bill ſerrated with ſharp indentings, the better to ſecure their prey; all of them almoſt are more voracious than the land birds, and there are ſome, as the

Ducks

Ducks and Gulls, &c. which devour indiſcriminately carrion and entrails.

We muſt divide the numerous claſs of Aquatic Birds into two great families: ſuch as ſwim and have palmated feet; and ſuch as haunt the ſhores and have divided feet*. The latter are differently ſhaped, their body being ſlender and tall, and as their feet are not webbed, they cannot dive or reſt on the water; they keep near the margin, and, wading with their tall legs among the ſhallows, they ſearch by means of their long neck and bill for their ſubſiſtence in the mud: they are a ſort of amphibious animals, that occupy the limits between the land and the water, and fill up the gradations in the ſcale of exiſtence.

Thus the aërial inhabitants conſiſt of three diviſions, which have each their ſeparate abode: ſome are appointed by nature to reſide on the land; others are deſtined to ſail on the water; and to an intermediate tribe, the confines of theſe two elements have been allotted: life has been varied in all its poſſible forms, and the immenſe richneſs of creation diſplayed to our admiration and aſtoniſhment.

* Ariſtotle, *Hiſt. Anim.* Lib. ix. 16.

We have often had occasion to remark that none of the quadrupeds and few of the birds which inhabit the Southern regions of the one continent, are found in the other; being unable to traverse the vast extent of intervening ocean. But this law entirely fails in the present instance; the Aquatic birds occur equally in the Old and in the New World, and even in the remotest islands of the habitable globe *. And this privilege is even extended to such as frequent only the shores: for, by tracing the line of coast, they may arrive at the extremities of both continents; nor, in their progress, will they experience much change of climate, since the heat is tempered and the cold mitigated, by the sea-breezes. Accordingly many species of shore birds, which in our continent retire to the North in summer, seem to have passed by degrees into the boreal tracts of America †

Most of these Aquatic birds appear to be half nocturnal ‡; the Herons roam during the night; the Woodcock only begins to fly in the evening; the Bittern still screams after the decline of day; the Cranes are heard to

* See the articles of the *Flamingos*, of the *Pelican*, of the *Frigat*, of the *Tropic Bird*, &c.

†. See the articles of the *Plovers*, of the *Herons*, of the *Spoon-bills*, &c.

‡ Edwards.

cry

cry aloft in the air, admidſt the ſilence and darkneſs of night; this alſo is the period when the Gulls range abroad, when the wild Geeſe and Ducks alight in our rivers, and ſtay uſually longer than during the day. Theſe habits are derived from ſeveral circumſtances connected with their ſupport and ſecurity: the coolneſs of the evening entices the worms to come out of their holes; the fiſhes are then in motion, and the general obſcurity conceals theſe birds from their enemies. Yet not always prudent in directing their attacks, they ſometimes fall victims of their own raſhneſs or impetuoſity. We have found a King fiſher in the belly of an Eel; the Pike often catches the birds that dive or glance over the ſurface, and even thoſe which come to drink or bathe at the margin of the pool; and, in the frozen ſeas, the Whales open their enormous jaws, to ſwallow not only whole columns of Herrings and other fiſhes, but alſo the birds which hover in purſuit of theſe, ſuch as the Albatroſſes, the Penguins, and the Scoter Ducks, &c.

Thus nature, while ſhe beſtows great privileges on the birds of the water, annexes alſo to their condition ſome inconveniences; and ſhe has even withheld from them one of her nobleſt gifts. None of them has the power of warbling, and what has been ſaid of the ſong of the Swan is altogether fabulous. The voice of the Aquatic Birds is ſtrong, harſh, and loud, calculated to be heard at a diſtance, and to reſound on the wide-ſpread ſhores: it conſiſts of raucous notes, of cries, and of ſcreams, and has none of thoſe flexible and ſoft accents, nor that ſweet melody with which our rural chanters enliven the grove, when they proclaim the delights of ſpring and of love: as if the formidable element, the ſcene of ſtorms, had for ever repelled theſe charming birds, whoſe peaceful ſong required days of ſerenity and nights of calm; as if the ocean permitted its winged inhabitants to utter nothing but coarſe and ſavage ſounds, which pierce through the horrors of the tempeſt, and are heard amidſt the roaring of the blaſt and the daſhing of the ſurge.

The number of Aquatic birds, including thoſe which haunt the ſhores, and reckoning the individuals, is perhaps equal to that of the Land-birds. If the latter are diſperſed through the hills and the vales, through the foreſts and the open fields: the former, plying by the edge of the water, or riding on the waves, inhabit a ſecond element as vaſt and as free as the air itſelf; their food, alſo,

alſo, is more abundant, and depends not on the caprice of the ſeaſons, or on the produce of human induſtry. Hence, too, the water birds aſſociate more habitually than the land birds, and form larger flocks; for inſtance, few ſpecies of theſe, at leaſt of an equal bulk, are ſo numerous, in the ſtate of nature, as the Geeſe and the Ducks. In general, there is greater union among animals, the further they are removed from the controul of man.

But both the ſpecies and the individuals of the land birds are more plenty in proportion as the climate is hotter; whereas the water birds ſeem to prefer the cold regions. Mariners inform us that the Herring-gulls, the Penguins, and the Scoter-ducks appear in myriads on the frozen ſhores of the North, as do the Albatroſſes, the Manchots, and the Petrels, on the frigid iſlets in the high ſouthern latitudes.

Yet the birds of the land ſeem to ſurpaſs thoſe of the water in fecundity; none of an equal ſize are ſo prolific as the gallinaceous tribe. Nor can this difference be attributed to the abundant and generous food which the domeſtic ſtate affords, for the tame Gooſe and Duck never lay ſo many eggs as the common Hen. Theſe Aquatic birds

birds are rather prisoners than domestics; they still retain the traces of their primæval liberty, and shew a degree of independence, which the land-fowl seem to have totally lost. They die if kept confined; they require to roam at large, and enjoy part of their natural freedom in the freshning pools. Nay, if their wings be not clipped, they will often join their wild brethren, and make their escape *. The Swan, that ornament of the artificial lakes in our superb gardens, sails along rather with the firm dignity of a master than the humble deportment of a slave.

As domestication imposes little constraint on the Aquatic birds, it introduces but slight alterations in their shape or plumage. The tame Duck admits of few varieties; while the Cock presents such a number of new breeds, that they seem almost to confound and obliterate the original stock. The birds of the water are also less known than those of the land; and, by placing

* Though there are instances of tame Ducks and Geese joining wild ones, they probably meet with harsh usage from their associates: for the antipathy between the domestic and the wild birds subsists in these species as in all others. The Sieur Irecourt, a person of veracity whom I have frequently cited, having put into a pond wild Ducks taken from the nest in a marsh, with other tame Ducks of nearly the same age, the latter assailed them, and in less than two or three days killed them outright.

them

them on the ocean, nature ſeems to have removed them from the empire of man.

The ſeas which abound moſt in fiſh, attract and eſtabliſh on their ſhores infinite multitudes of Aquatic birds. Innumerable flocks inhabit the Sambal iſlands and the coaſt of the iſthmus of Panama, particularly on the Northern ſide; nor is the Weſtern ſide of the continent leſs frequented on the Southern coaſt, but there are few on the Northern. Wafer aſſigns for the reaſon, that in the bay of Panama the fiſhes are not ſo plenty as at the Sambal iſlands. The great rivers of North America are entirely covered with water birds. The ſettlers at New Orleans, who kill them on the Miſſiſſippi, formed a ſmall branch of commerce in the fat or oil extracted from them. Many iſlands have been called *Bird Iſlands*, being deſert and wholly overſpread with ſea-fowl. The iſland of *Aves* among others, fifty leagues to the leeward of Dominica is ſettled by unequaled numbers. There we find Plovers, Red-ſhanks, Gallinules, Flamingos, Pelicans, Gulls, Frigats, Boobies, &c. Labat, who publiſhes theſe facts, remarks, that this coaſt is exceedingly rich in fiſh, and that the high-water mark is conſtantly covered by an immenſe quantity of ſhells.

The

The fiſh ſpawn, alſo, which often floats on the ſurface of the ſea near the great banks, attracts equally the birds *. There are certain parts on the coaſts and in the iſlands, where the whole ſoil, to a conſiderable depth, conſiſts entirely of the excrements of water fowls; ſuch is the caſe near the Peruvian coaſt, on the iſland of Iquique, whence the Spaniards carry the dung to manure their lands on the continent †. The rocks of Greenland are covered to their tops with a ſort of turf compoſed of the ſame ſubſtance, and the relics of old neſts ‡ Theſe birds are numerous alſo on the rocks of the Norwegian ſhore §, and on the

* "In the 41ſt degree of South latitude near Chili, we met on the ſurface of the ſea with a bed of fiſh-ſpawn, which extended about a league, and as we had obſerved another bed the day before, we judged what it was that had attracted the birds which we had ſeen for two or three days." *Obſervations du P. Feuillée (edit. 1725.) p.* 79.

† For more than a century paſt ſeveral ſhips have been annually loaded with this dung reduced to mould, which the Spaniards call *guana*, and carry it to fertilize the neighbouring vallies, particularly that of Arica, which by the aſſiſtance of this manure, bears the pimento. See *Frezier* and *Feuillée*. "From Cape Horn we ſteered without the rocks that lie off Miſtaken Cape. Theſe rocks are white with the dung of fowls, and vaſt numbers were ſeen about them." *Cook's Second Voyage.* Vol. ii. p. 190.

‡ See *Hiſt. Gen. des Voyages* tom. 19. p. 27.

§ The Aquatic birds of the coaſts of Norway are common alſo to Iceland and Ferol; they are ſo numerous that the inhabitants

the islands of || Iceland, and Feroe, where the eggs are the principal support of the inhabitants, who gather them in the precipices and most frightful cliffs *. Such also

inhabitants live on their flesh and their eggs. They fatten the country with their dung, and their feathers afford a considerable branch of trade to the town of Berger. *Pontoppidan's Natural History of Norway.*

|| The sea-fowl appear in vast flocks on the islets near Iceland, and spread to the distance of twelve or fifteen leagues: the sight of these betokens the approach to the island. Among these birds, are different kinds of Gulls, most of them described in Marten's Voyage to Spitzbergen. *Horrebow's Description of Iceland.*

* "The birds which stock the coasts of Iceland, seek to make their nests in the most inaccessible places and on the most craggy rocks: however, the inhabitants can plunder these, notwithstanding the danger of the pursuit. I have myself," says Horrebow, "seen the manner of taking them, and I must confess that I never could behold, without shuddering, the intrepidity with which the men risked their lives: several of these people have fallen into the sea, or been dashed against the precipices over which they were obliged to be suspended. On the top of the rock is fastened as firmly as possible, a beam that projects a considerable way: this bears a pulley and a rope, by means of which a man tied by the middle of the body descends along the rocks: he holds a long pole armed with an iron hook, by which he guides himself among the crags. Upon a certain signal being made, the men who are stationed on the summit of the precipice draw him up with his plunder of one or two hundred eggs. The search is continued as long as any eggs can be found, or as long as it is possible to endure this suspension, which becomes very fatiguing. During this employment, the birds are seen flying in thousands, uttering frightful screams. The inhabitants of those places reap great profit from this species of industry; for besides the eggs,

alſo are the deſert and almoſt inacceſſible iſlets in the Barra Firth, near the coaſt of Scotland, which are viſited annually by the people from Hirta, who collect the eggs by thouſands and kill the birds †. And laſtly, they are ſo plentiful in the Greenland ſeas, that in the language of the country there is a word to expreſs the method of fowling

eggs, they find many young birds, of which ſome ſerve as food, and others yield abundance of feathers, that are ſold to the Daniſh merchants." *Horrebow.*—Pontoppidan deſcribes the ſearch for eggs which is made in Norway, as no leſs frightful. "The cavities where theſe birds breed occur in the craggy perpendicular rocks all along the coaſt. To climb to them, a perſon puts a rope round his body, while his companions puſh againſt his back with a long pole, to help him up to ſome place where he can reſt his foot and faſten his rope; then they withdraw the pole and a ſecond clambers up in the ſame manner. After they have joined, they both tie themſelves to the ſame rope, and aſſiſt each other in mounting higher by means of an iron hook, puſhing and drawing up each other by turns. The birds ſuffer themſelves to be caught by the hand on their neſts in the caverns, and the ſpoils are thrown to thoſe who wait below the rock in a boat. Theſe fowlers are ſometimes eight days abſent from their companions, and often tumble together into the ſea. When they want to enter the hollows of mountains, the boldeſt is let down by a rope from the top of the rock. He wears a large ſtrong hat, to ſcreen him from the blows of the ſtones that may fall: when he would enter any cavity, he preſſes his feet againſt the mountain, puſhes back with all his force, and directs ſo well his body that he lands ſtraight in." *Hiſt. Nat. de Norwege*, par Pontoppidan, *part ii. Journal Etranger, mois de fevrier*, 1757.

† See Collection of different treatiſes on Phyſics and Natural Hiſtory, by *M. Deſlandes*, tom. i. p. 163.

by

by chaſing the birds into the narrow inlets, where, being hemmed in, they are taken in vaſt numbers §.

The Water-fowl are alſo the inhabitants which nature has aſſigned to the diſtant iſlands that are loſt in the midſt of an immenſe ocean, whither the other ſpecies which live on the ſurface of the land could never have penetrated *. Navigators have found birds in poſſeſſion of thoſe ſolitary and inhoſpitable ſpots which ſeemed unfit for the abode of animated beings †. They are ſpread from North to South ‡, but no where

§ Sarpſipock. *Dict. Groënland. Hafniæ.*

* "Scarce had the veſſel anchored (at the iſland of Aſcenſion) than thouſands of birds perched on the maſts and rigging; the fall of five hundred, which were killed in the ſpace of a quarter of an hour, did not deter the reſt from flying about the ſhip; they became ſo importunate that they bit the hats and caps of twenty men who went aſhore." *Relation de Rennefort. Hiſt. Gen. des Voyages, tom. viii. p.* 583,

† "We obſerved theſe rocks (at Eaſter-iſland,) whoſe cavernous aſpect and black ferruginous colour, bore the marks of a ſubterraneous fire. We remarked two in particular, the one like an enormous column or obeliſk, and both filled with innumerable Sea-fowl, whoſe diſcordant cries ſtunned our ears. *Forſter.*

‡ I went on ſhore (at Port Deſire in the Straits of Magellan), the river as far as I could ſee, was very broad; there were in it a number of iſlands, ſome of which were very large, and I make no doubt but that it penetrates the country for ſome hundred miles. It was upon one of theſe iſlands that I went on ſhore, and I found there ſuch a number of birds

where more numerous than in the Frigid Zones §; becauſe in thoſe dreary regions, the naked earth bound in icy torpor, is blaſted with perpetual ſterility; while the ſea yet teems with life || Accordingly, navigators and naturaliſts have remarked, that in the arctic countries there are few Land-birds in compariſon of Water birds *:

birds, that when they roſe they literally darkened the air, and we could not walk a ſtep without treading upon their eggs." *Byron's Voyage in Hawkeſworth's Collection. Vol.* 1, *p.* 21.

§ Gmelin ſays, that he never ſaw ſuch a large number of birds aſſembled in flocks as at Mangaſea on the Jeniſea: It was in the month of June; the moſt numerous were the Water fowl, Geeſe of all kinds, Ducks, Gallinules, Gulls, and the Shore-birds, Woodcocks, Divers, &c. *Hiſt. Gen. des Voyages. tom. xviii. p.* 357.

|| The Albatroſſes now left us during our paſſage amidſt the iſlands of ice; and we ſaw only one from time to time. The Tropic-birds, the Cut-waters, the little Grey birds, the Swallows, were no longer ſo numerous. On the other hand, the Penguins began to appear, for to-day we ſaw two. Notwithſtanding the coldneſs of the climate we obſerved the White Petrel conſtantly among the maſſes of ice, and we may regard it as a certain fore-runner of ice. From its colour we took it for the Snowy Petrel. Many Whales ſhewed themſelves among the ice, and varied a little the frightful ſcene. We paſſed more than eighteen iſlands of ice, and ſaw New Penguins." *Captain Cook's Second Voyage.*

* See the *Fauna Suecica* of Linnæus, the *Ornithologia Borealis* of Brunnich, the *Zoologia Danica* of Muller. The ſame obſervation holds with regard to the Antarctic regions." "Very few Land-birds are found in Terra del Fuego: Mr. Banks ſaw none larger than our Black-birds; but there was great abundance of Water-fowl, particularly Ducks." *Cook's Firſt Voyage.*

the

the former require herbs, ſeeds and fruits, of which the ground yields only a few ſtunted ſpecies; while the latter ſeek the land only as a place of refuge, a retreat in tempeſts, a ſtation in dark nights, and a ſupport for their neſts. Cook and Forſter, in their voyages to the South Sea, ſaw many of theſe birds reſting and ſleeping on the floating ice *; and ſome of them even breed on the ice †. Indeed what can be colder or harder than their uſual beds on the frozen ſummits ‡.

This laſt fact proves that the Water-fowls are the remoteſt inhabitants of our globe, and are well acquainted with the Polar regions. They penetrate into lands where the White Bear no more appears;

* See the articles of the Petrels and Penguins.

† "We met with a great bank of ice, on which we were obliged to moor (at Nova Zembla); ſome ſailors mounted upon it and gave a very ſtrange account of its figure, it was all covered with earth to the top, and they found on it near forty eggs." *Relation de Hemſkerke and Barentz. Hiſt. Gen. des Voy. tom. xvi. p.* 112.

‡ On the 22d July, being near Cape Cant (at Nova Zembla) we went repeatedly aſhore to ſeek for birds eggs: there was plenty of neſts, but in ſteep precipices; the birds ſeemed not to be afraid of the ſight of men, and moſt of them ſuffered themſelves to be caught by the hand. Each neſt had but a ſingle egg, on the bare rock without ſtraw or feathers to keep it warm: a ſight which aſtoniſhed the Hollanders, who could not conceive how eggs were covered and hatched in ſuch intenſe cold." *Id. Ibid.*

 into

into ſeas abandoned by the Seal, the Walrus, and the other amphibious animals. They live agreeably in thoſe climates during the whole ſummer, and retire about the autumnal equinox, when the night encroaches faſt on the day, at laſt totally extinguiſhes it, and wraps the awful ſcene in tedious darkneſs. They ſpend the ſhort winter-days in lower latitudes, and return again in the ſpring to their frozen abodes.

No. 17.3

THE STORK.

The STORK.

La Cigogne. Buff.
Ardea-Ciconia. Linn. and Gmel.
Ciconia. All the naturaliſts.
The White Stork. Penn and Lath *.

WE have obſerved that between the land birds and the ſea birds, which have webbed feet and reſt on the ſurface, there is a large claſs that haunt the ſhores, and, being furniſhed with toes, are deſtined to tread the ground, but at the ſame time are enabled, by their long neck and bill, to find their food under the liquid element. Of the numerous families which frequent the ſides of rivers and the ſea-beach, that of the Stork, which is the beſt known and the moſt celebrated, occurs firſt: it contains two ſpecies, the white and the black, which are exactly of the ſame form, and have no external difference but that of colour. This diſtinction might be totally

* In Greek Πελαργος: in Latin *Ciconia*: in Italian *Cigogna*: in Spaniſh *Ciguenna*: in Hebrew and Perſian *Chaſida*: in Chaldean *Chavarita*, *Deiutha* and *Macuarta*: in Arabic *Zakid*, according to Geſner, and *Leklek*, according to Dr. Shaw: in Mooriſh *Bell-Arje*: in Poliſh *Bocian Czarni*, *Bocian Snidi*: in Flemiſh *Ouweaer*: in German *Storck*.

 diſregarded,

disregarded, were not their instincts and habits widely different. The Black Stork prefers desert tracts, perches on trees, haunts unfrequented marshes, and breeds in the heart of forests. The White Stork, on the contrary, settles beside our dwellings; inhabits towers, chimnies, and ruins: the friend of man, it shares his habitations, and even his domain; it fishes in our rivers, pursues its prey into our gardens, takes up its abode in the midst of cities, without being disturbed by the noise and bustle *, and ever respected and welcomed, it repays, by its services, the favours bestowed on it: as it is more civilized, it is also more prolific, more numerous, and more dispersed, than the Black Stork, which appears confined to particular countries, and resides always in the most sequestered spots.

The White Stork is smaller than the Crane, but larger than the Heron; its length, from the point of the bill to the end of the tail, is three feet and an half, and to the nails, four feet; the bill, from the tip to the corners, measures seven inches; the leg eight inches; the naked part of the thighs five, and the extent of the

* Witness the Stork's nest built on the temple of Concord in the Capitol, and mentioned by Juvenal, *Sat.* I. *ver.* 116, and which is also represented in the medals of Adrian.

wings

wings is more than ſix feet. It is eaſy to form an idea of it; its body is of a bright white, and its wings black, characters which its Greek name expreſſes †; its legs and bill are red, and its long neck is arched: theſe are the obvious features; but on cloſer examination, we perceive, on the wings, violet reflections and ſome brown tints, we may count thirty quills in the wing, when it is ſpread; they form a double ſcalloping, thoſe next the body being almoſt as long as the outer ones, and equal to them when the wing is cloſed: in that ſituation, the wings cover the tail; and when they are expanded for flying, the great quills ſhew a ſingular diſpoſition; the firſt eight or nine part from each other, and appear diverging and detached, ſo that a ſpace is left between each, a property to be found in no other bird: the feathers below the neck are white, longiſh, and pendulous; in which reſpect, the Storks reſemble the Herons, but their neck is ſhorter and thicker: the orbits are naked, and covered with a wrinkled ſkin of reddiſh black; the feet are covered with ſcales in hexagonal tablets, and broader the higher they are placed. There are rudiments of membranes

† Πελαργος from Πελος black, and αργος white.

between the great toe and the inner toe as far as the firſt joint, which, projecting on the outer toe, ſeem to form the gradation by which nature paſſes from the birds that have the feet parted by toes, to thoſe that have them webbed: the nails are blunt, broad, flat, and much like the human nails.

The Stork flies ſteadily and with vigour, like all the birds furniſhed with broad wings and a ſhort tail: it holds its head ſtraight forward, and ſtretches back its legs, to directs its motion ‡: it ſoars to a vaſt height, and performs diſtant journies, even in tempeſtuous ſeaſons. The Storks arrive in Germany about the eighth or tenth of May §, and are ſeen before that time in the provinces of France. Geſner ſays, that they precede the Swallows, and enter Switzerland in the month of April, and ſometimes earlier. They arrive in Alſace in March, or even in the end of February. Their return is ever auſpicious, as it announces the ſpring. They inſtantly indulge thoſe tender emotions which that ſeaſon inſpires: Aldrovandus paints with warmth their mutual ſigns of felicity and love, the eager congratulations, and the fondling endearments of the male and female, when they

‡ Ariſtotle, *Lib. ii.* 15.

§ Klein, *De avibus errat. & migrat.*

arrive

arrive at their neſt after their diſtant journey ||: for the Storks always ſettle in the ſame ſpots, and, if their neſt has been deſtroyed, they rebuild it with twigs and aquatic plants, and uſually on lofty ruins, or on the battlements of towers, and ſometimes on large trees beſide water, or on the point of bold cliffs. In France, it was cuſtomary in Belon's time, to place wheels on the houſe-tops, to entice the Stork to neſtle. This practice ſtill ſubſiſts in Germany and in Alſace: and in Holland, ſquare boxes are planted on the ridge, with the ſame view *.

When the Stork is in a ſtill poſture it reſts on one foot, folds back its neck, and reclines its head on its ſhoulder. It watches the motions of reptiles with a keen eye, and commonly preys on frogs, lizards, ſerpents, and ſmall fiſh, which it finds in

|| "When they have arrived at their neſt .... good God! what ſweet ſalutation; what gratulation for their proſperous return! what embraces! what honied kiſſes! what gentle murmurs they breathe!" *Tom. iii. p.* 298.

* Lady Montague in her Letters, No. 32. ſays that the Storks neſtle on the ground in the ſtreets. If ſhe is not miſtaken with regard to the ſpecies of theſe birds, the protection which the Stork enjoys in Turkey muſt have ſingularly emboldened it; for in our countries, it always chuſes the moſt innacceſſible places, which may command the vicinity, and conceal it in the neſt:

marſhes, by the ſides of the ſtreams, and in wet vales.

It walks like the Crane with long meaſured ſtrides. When it is irritated or diſcompoſed, or even actuated by the amorous paſſions, it makes with its bill, a repeated clattering, which the ancients expreſs by the ſignificant words *crepitat*, *glotterat* †, and which Petronius accurately marks, by the epithet *crotaliſtria* ‡, formed from *crotalum*, the caſtanet or rattle. In this ſtate of agitation it bends its head back in ſuch a manner, that the lower mandible appears uppermoſt, and that the bill lies almoſt parallel on the back; and in this attitude, the two mandibles ſtrike violently againſt each other; but in proportion as it raiſes up its neck, the clattering abates, and ceaſes when the bird has reſumed its ordinary poſture. Such is the only noiſe which the Stork ever makes, and, as it ſeems dumb, the ancients were probably induced to ſuppoſe that it had no tongue §: this, indeed, is ſhort and concealed in the entrance of the throat, as in all the birds with long bills, which have alſo a particular

† *Quæque ſalutato crepitat concordia nido.* Juvenal, Sat. I. *Glotterat immenſo de turre ciconia roſtro.* Aut. Philomel.

‡ Publius Syrus had made the ſame application of this word.

§ Pliny, *Lib*, *x*. 31.

mode

mode of ſwallowing, they by a certain caſt of the head, toſſing their food into the throat. Ariſtotle makes another remark with regard to birds which have long necks and bills, that their excrements are always thinner than thoſe of other birds ||.

The Stork does not lay more than four eggs, oftner not more than two; they are of a dirty and yellowiſh white, rather ſmaller, but longer, than thoſe of a Gooſe. The male ſits when the female goes in queſt of food; the incubation laſts a month; both parents are exceedingly attentive in bringing proviſions to the young, which riſe up to receive it, and make a ſort of whiſtling noiſe *. The male and female never leave the neſt at once; but, while the one is employed in ſearching for its prey, the other ſtands near the ſpot on one leg, and keeps an eye conſtantly on the brood. When firſt hatched, the young are covered with a brown down, and their long

|| *Hiſt. Anim.* Lib. ii. 22.

* Ælian ſays that the Stork vomits food to its young, which muſt not be underſtood of aliments partly digeſted, but of recent prey which it diſgorges from its æſophagus, or even from its ſtomach whoſe aperture is ſufficiently large. See the obſervation of Peyerus *de ciconiæ ventre & affinitate quadam cum ruminantibus. Ephem. Nat. curios. dec.* 2. *ann.* 2. *obſ.* 97. See alſo two anatomical deſcriptions of the Stork, the one by Schelhammer, *Collect. Acad. part etrang.* vol. iv. obſ. 109, and the other by Olaus Jacobæus. *Id. obſ.* 94.

long ſlender legs not having yet ſtrength enough to ſupport them, they creep upon their knees †. When their wings begin to grow, they eſſay their force in fluttering about the neſt; though it often happens, that, in this exerciſe, ſome of them fall, and are unable to regain their lodgement. After they venture to commit themſelves to the air, the mother leads them, and exerciſes them in ſmall circumvolutions about the neſt, whither ſhe conducts them back. And about the latter end of Auguſt, the young Storks having now attained ſtrength, join the adults, and prepare for migration. The Greeks have placed their rendezvous in a plain of Aſia, called the *Serpent's Diſtrict*, where they aſſembled ‡ as they do now in ſome parts of the Levant §, and even in Europe, as in Brandenburg and elſewhere.

† Obſervation of Biſhop Grunner. *Mem. Soc. of Drontheim.*

‡ *Pythonos comen, quaſi ſerpentium pagum, vocant in Aſiâ, patentibus campis, ubi congregatæ inter ſe commurmurant, eamque quæ noviſſima advenit lacerant, atque ita abeunt. Notatum poſt idus auguſtas non temere viſas ibi.* Plin. Lib. x. 31.

From this paſſage it appears that the aſſembly of the Storks is not without tumult and even fighting; but that *they tear the laſt comer,* as Pliny aſſerts, is doubt a fable.

§ "It is remarked that the Storks before they paſs from one country into another, aſſemble a fortnight beforehand, from all the neighbouring parts, in a plain, holding once a day a *divan*, as they ſay in that country, as if their object was to fix the preciſe time of their departure and the place of their retreat." *Shaw's Travels.*

When

When they are convened previous to their departure, they make a frequent clattering with their bill, and the whole flock is in tumultuary commotion; all ſeem eager to form acquaintance, and to conſult on their projected rout, of which the ſignal in our climate is the North wind. Then, the vaſt body riſes at once, and, in a few ſeconds, is loſt in the air. Klein relates, that, having been called to witneſs this ſight, he was a moment too late, and that the whole flock had already diſappeared. Indeed, this departure is the more difficult to obſerve, as it is conducted in ſilence ||, and often during the night *. It is aſſerted, that in their paſſage, before they venture to croſs the Mediterranean, the Storks alight in great numbers in the neighbourhood of Aix † in Provence. Their departure appears to be later in warm countries; for Pliny ſays, that *after the retreat of the Stork, it is improper to ſow* ‡.

|| Belon ſays, that it is not remarked, becauſe they fly without noiſe or cries, while the Cranes and Wild-geeſe, on the contrary, ſcream much on the wing.

* *Nemo vidit agmen diſcedentium, cum diſceſſurum appareat; nec venire, ſed veniſſe cernimus; utrumque nocturnis fit temporibus.* Pliny, *Lib. x.* 31.

† Aldrovandus.

‡ *Poſt ciconiæ diſceſſum male ſeri.* Lib. viii. 41.

Though

Though the ancients had obſerved the migrations of the Storks, they were ignorant of the countries to which they retired §. Some modern travellers have made good obſervations on that ſubject: in autumn the plains of Egypt are entirely covered with theſe birds. "It is perfectly aſcertained, ſays Belon, that the Storks winter in Egypt, and in Africa; for we have ſeen the plains of Egypt whitened by them in the months of September and October. At that ſeaſon, when the waters of the Nile have ſubſided, they obtain abundance of food; but the exceſſive heats of ſummer drive them to more temperate climates; and they return again in winter, to avoid the ſeverity of the cold; the contrary is the caſe with the Cranes, which viſit us with the Geeſe in winter, when the Storks leave us." This remarkable difference is owing to that of the climates which theſe birds inhabit; the Geeſe and Ducks come from the North, to eſcape the rigors of the winter; the Storks leave the South, to avoid the ſcorching heats of ſummer ||.

Belon

§ Jeremiah viii. 7.

|| Several authors pretend that the Storks do not retire in winter, but then lurk in caverns, or even at the bottom of lakes. This was the common opinion in the time of Albertus Magnus. Klein relates, that two Storks were dragged out

Belon ſays alſo, that he ſaw them wintering round Mount Amanus near Antioch, and paſſing about the end of Auguſt towards Abydus, in flocks of three or four thouſand, from Ruſſia and Tartary. They croſs the Helleſpont; and on the ſummits of Tenedos, they divide into ſquadrons, and diſperſe themſelves Northwards.

Dr. Shaw ſaw at the foot of Mount-Carmel, a flight of Storks from Egypt to Aſia; about the middle of May 1722. "Our veſſel, ſays that traveller, being anchored under Mount Carmel, I ſaw three flocks of Storks, each of which was more than three hours in paſſing, and extended a half mile in breadth †." Maillet ſays,

out of the water in the pools near Elbing (*De avibus errat. & migrat. ad calcem.*) Gervais of Tillebury (*Epiſt. ad Othon iv.*) ſpeaks of other Storks that were found cluſtered in a lake near Arles; Merula in Aldrovandus ſpeak of thoſe which fiſhermen drew out of the lake of Como; and Fulgoſus, of others that were fiſhed near Metz (*Memorab. lib. i. cap. 6.*) Martin Schoockius, who wrote a ſmall treatiſe on the Stork, printed at Groningen in 1648, ſupports theſe teſtimonies. But the hiſtory of the migrations of the Storks is too well known, not to attribute to accidents the facts juſt mentioned, if they indeed may be relied on. See further the Article of the Swallow.

† He adds; "Theſe Storks came from Egypt, becauſe the channel of the Nile and the marſhes which it makes annually, being dried, they retire to the North Eaſt." But this author is miſtaken; the Storks rather flee from the inundation which covers the whole country; the river having no banks after the end of April.

that

that he ſaw the Storks deſcend towards the end of April from Upper Egypt, and halt on the grounds of the Delta, which the inundation of the Nile ſoon obliges them to leave ‡.

Theſe birds which thus remove from climate to climate, never experience the rigors of winter; their year conſiſts of two ſummers, and twice they taſte the pleaſures of the ſeaſon of love. This is a remarkable peculiarity of their hiſtory, and Belon poſitively aſſures us that the Stork has its ſecond brood in Egypt.

It is ſaid, that Storks are never ſeen in England, unleſs they are driven upon the iſland by ſome ſtorm. Albin remarks as a ſingular circumſtance, that there were two of theſe birds at Edgware in Middleſex, and Willughby declares, that the figure which he gives was deſigned from one ſent from the coaſt of Norfolk, where it had accidently dropped. Nor does the Stork occur in Scotland, if we judge from the ſilence of Sibbald. Yet it often penetrates into the Northern countries of Europe;

‡ Some Crows intermingle at times with the Storks in their paſſage, which has given riſe to the opinion of St. Baſilius and Iſidòrus, that the Crows ſerve to direct and eſcort the Storks. The ancients have alſo ſpoken much of the combats between the Storks and the Ravens, the Jays, and other ſpecies of birds, when their flocks returning from Lybia and Egypt, met about Lycia and the river Xanthus.

rope; it is found in Sweden, according to Linnæus, and over the whole of Scania, in Denmark, Siberia, at Mangasea on the river Jenisca, and as far as the territories of the Jakutes §. Great numbers of Storks are seen also in Hungary ||, in Poland and Lithuania *; they are met with in Turkey, and in Persia, where Bruyn observed their nest carved on the ruins of Persepolis; and according to that author, they are dispersed through the whole of Asia, except the desert parts, which they seem to shun, and the arid tracts, where they cannot subsist.

Aldrovandus assures us, that Storks are never found in the territory of Bologna; they are rare even through the whole of Italy, where Willughby during a residence of twenty-eight years saw them only once, and where Aldrovandus owns he never saw them. Yet it appears, from Pliny and Varro, that anciently they were common; and we can hardly doubt, but that in their rout from Germany to Africa, or in their return, they must pass over Italy and the islands of the Mediterranean. Koempfer says, that the Storks reside the whole year in Japan; that would be the only country where they are stationary; in all others, they retire a few months after their arrival.

§ Gmelin. *Hist. Gen. des Voy. tom. xviii. p.* 300.

|| Marsigli.

* Klein.

In

In France, Lorraine and Alſace, are the provinces where theſe birds are the moſt numerous; there they breed, and few towns or villages in Lower Alſace are without Storks' neſts on their belfries.

The Stork is of a mild diſpoſition, neither ſhy nor ſavage; it is eaſily tamed; and may be trained to reſide in our gardens, which it will clear of inſects and reptiles. It ſeems to have an idea of cleanlineſs, for it ſeeks the by-corners to lay its excrements. It has almoſt always a grave air, and a mournful viſage; yet, when rouſed by example, it ſhews a certain degree of gaiety, for it joins the frolics of children, hopping and playing with them *. In the domeſtic condition it lives to a great age, and endures the ſeverities of our winters †.

To this bird are aſcribed moral virtues, whoſe image is ever venerable; temperance,

* "I ſaw in a garden, where the children were playing at hide and ſeek, a tame Stork join the party, run its turn when touched, and diſtinguiſh the child, whoſe turn it was to purſue the reſt, ſo well as to be on its guard." *Notes on the Stork by Dr. Hermann of Straſburg.*

† Ger. Nic. Heerkens of Groningen, who has written a ſmall Latin poem on the Stork, ſays that he kept one fifteen years, and ſpeaks of another which lived twenty-one years in the Fiſh-market of Amſterdam, and was interred with ſolemnity by the people. See alſo the obſervation of Olaus Borrichius on a Stork aged more than twenty-two years, and which became gouty. *Collect. acad. part. etran. tom. iv. p.* 331.

conjugal

conjugal fidelity ‡, filial and paternal piety §. It is true, that the Stork beftows much time on the education of its young, and does not leave them till they have ftrength fufficient for their defence and fupport; that when they begin to flutter out of the neft, the mother bears them on her wings; that fhe protects them from danger, and fometimes perifhes with them rather than forfake them ||. The Stork fhews tokens of attachment to its old haunts, and even gratitude to the perfons who have treated it with kindnefs. I am affured, that it has been heard to rap at the door in paffing, as if to tell its arrival, and give a like fign of adieu on its departure *. But thefe moral qualities are nothing in comparifon of the affection and tender offices which thefe birds lavifh on their

‡ "A great number of Storks neftle and breed in the neighbourhood of Smyrna. The inhabitants amufe themfelves with putting Hens' eggs into a Stork's neft: when the Chicks are hatched, the male Stork, feeing thefe ftrange figures makes a frightful noife, and thus attracts a multitude of other Storks, which peck the female to death, while the male vents lamentable fcreams." *Annual Regifter for* 1768.

§ Hence Petronius ftyles it *pietatis cultrix*.

|| See, in *Hadrianus Junius*, (*Annal. Battav. ad ann.* 1535,) the hiftory, famous in Holland, of the Delft Stork which in the conflagration of that city, after having in vain attempted to refcue her young, perifhed with them in the flames.

* Aldrovandus.

aged

aged and infirm parents †. The young and vigorous Storks frequently carry food to the others, which resting on the brink of the nest seem languid and exhausted, whether hurt by some accident, or worn out by years, as the ancients assert, nature having implanted in brutes that venerable piety, as an example to man, in whose breast the delicious sentiment is too often obliterated. The law which compelled the maintenance of parents was enacted in honour of them, and inscribed by their name. Aristophanes draws from their conduct a bitter satyr on the human race.

Ælian tells us, that the moral qualities of the Stork were the chief cause of the respect and veneration which it enjoyed among the Egyptians *; and the notion which

† Aristotle, *Hist. Anim.* Lib. ix. 20.

Ciconiæ senes, impotes volandi, nido se continent, ex his prognatæ terrâ marique volitant, & cibos parentibus afferunt, sic illæ, ut earum ætate dignum est, quiete fruuntur & copiâ; juniores vero laborem solantur pietate, ac spe recipiendæ in senectute gratiæ. Philo.

Genitricum senectam invicem alunt. Plin. *Lib. x.* 31.

See Plutarch, and all the ancients cited by Plutarch.

* Alexander the Myndian, in Ælian, says that the Storks worn out with old age, repair to certain islands in the ocean, where, in reward of their piety, they are changed into men. In auguries the appearance of the Stork denoted union and concord. (*Alexand. ab Alex. genial. dies*); its departure in the

which the vulgar ſtill entertain, that its ſettling on a houſe betokens proſperity, is perhaps a veſtige of the ancient opinion.

Among the ancients, it was held a crime to kill the Stork. In Theſſaly the murder of one of theſe birds was puniſhed by death; ſo precious were they held in that country, which they cleared of Serpents *. A portion of that regard is ſtill retained in the Levant †. The Stork was not eaten among the Romans, and a perſon who from a ſtrange ſort of luxury ordered it to be brought to his table, drew upon himſelf the obloquy of the people ‡. Nor is the fleſh recommended

the time of public calamity was regarded as a diſmal preſage. Paul the Deacon ſays, that Attila, having purpoſed to raiſe the ſiege of Aquileia, was determined to renew his operations, upon ſeeing Storks retiring from the city and leading away their young (*Æneas Sylvius Epiſt. ii.*) In hieroglyphics it ſignified piety and beneficence, virtues which its name expreſſed in the moſt ancient languages (*chaſida*, in Hebrew, according to Bochart); and we often ſee the emblem, as on the two beautiful medals of L. Antonius, given in Fulvius Urſinus, and in two others of Q. Metellus, ſurnamed *the Pius*, as reported by Paterculus.

* Pliny, Lib, x. 31.

† "The Mahometans have a great eſteem and veneration for the Stork, which they call *Bel arje*; it is almoſt as ſacred among them as the Ibis was among the Egyptians, and they would look upon a perſon as profane, who ſhould kill or even harm it." *Shaw's Travels.*

‡ As this ancient epigram atteſts:

Ciconiarum Rufus iſte conditor
Plancis duobus eſt hic elegantior.

Suffragiorum

mended by its quality §; and this bird formed by nature our friend and almoſt our domeſtic, was never deſtined to be our victim. [A]

Suffragiorum puncta ſeptem non tulit.
Ciconiarum populus mortem ultus eſt.

§ *Cornelius Nepos, qui divi Auguſti principatu obiit, cum ſcriberet turdos paulo ante cœptos ſaginari, addidit, ciconias magis placere quam grues; cum hæc nunc ales inter primarias expetatur, illam nemo velit attigiſſe.* Plin. Lib. x.

[A] Specific character of the White Stork, *Ardea Ciconia*: It is white: its orbits and wing-quills, black; its bill, its legs and its ſkin, blood coloured.

The BLACK STORK.

Ciconia Nigra. Linn. Gmel. &c. &c.
* *Cicouia Fusca.* Ray. Will. and Klein.

THOUGH, in all languages this bird is termed the *Black Stork*, it derives its epithet rather from the opposition to the White Stork, than from the dye of its plumage, which is generally brown mixed with fine changeable colours.

The back, the rump, the shoulders. and the coverts of the wings, are of a brown, which varies with violet and gold green; the breast, the belly, and the thighs, are cloathed with white feathers, and of the same colour are the coverts under the tail, which consists of twelve brown quills changing with violet and green; the wing contains thirty quills, which are of a varying brown, the green predominating in the first ten, and the violet in the remaining twenty; the feathers at the origin of the neck are of a brown glossed with violet, and washed with grayish at the tips: the throat and neck are covered with small

* In Italian *Aghiron Nero*. in German *Schwartze Storck.*

brown feathers, terminated by a whitiſh point; but this character is wanting in ſome individuals; the top of the head is brown, mixed with a violet gloſs and gold green: the eye is encircled by a very red ſkin, the bill, too, is red, and the naked part of the thighs, the legs, and the nails, are of the ſame colour; yet that property ſeems to admit ſome variety, for ſeveral naturaliſts, as Willughby, make the bill to be greeniſh like the legs: it is only a ſlight degree ſmaller than the White Stork, its alar extent being five feet ſix inches.

Savage and ſolitary, the Black Stork ſhuns our habitations, and haunts only the deſert fens; it neſtles in the heart of the woods, on old trees, eſpecially on lofty pines. It is common in the Swiſs Alps; it is ſeen by the edge of the lakes, watching its prey, flying on the ſurface of the water, and ſometimes diving haſtily for fiſh: yet it does not depend on that mode of ſubſiſting only; it gathers inſects among the herbage and the mountain dales; veſtiges of caterpillars and graſshoppers are found in its ſtomach. When Pliny ſays, that the Ibis occurs in the Alps, he miſtook the Black Stork for that Egyptian bird.

It is found in Poland *, Pruſſia, and Lithuania †, in Sileſia ‡, and in many other parts of Germany §: it penetrates as far as Sweden, always ſeeking the wild fenny tracts. How ſavage ſoever it appear, it may be kept in confinement, and even in ſome degree tamed. Klein aſſures us, that he fed one ſome years in his garden. We are not informed whether it migrates like the White Stork, nor whether the ſeaſons of its paſſage are the ſame. Yet there is every reaſon to entertain that opinion, for even in our climates there could be no proviſion for it in winter.

This ſpecies is not ſo numerous, or ſo widely diſperſed, as that of the White Stork; it ſeldom ſettles in the ſame places ‖, but ſeems to occupy the countries which the other neglects. Wormius, while he remarks that the Black Stork is very frequent in Sweden, adds that it is exceedingly rare in Holland, where the White Storks are known to be very numerous. Yet the Black Stork is not ſo rare in Italy

* Rzaczynſki.

† Klein.

‡ Schwenckfeld.

§ Willughby. It is very rare in all theſe countries.

‖ The Brown Stork only paſſes through Lorraine and does not halt. *Note communicated by M. Lottinger.*

as the white one *; and according to Willughby it is frequently seen with other Marsh-birds in the markets of Rome, though its flesh has an unpleasant fishy taste and a rank smell [A]

* Jo. Lincæus.

[A] Specific character of the Black Stork, *Ardea Nigra:* "It is brown; its breast and belly white." These birds soar to a vast height. Great flocks of them pass in the spring over Sweden, and stretch Northward. They make no halt in that country. They return to the South in autumn.

Foreign Birds which are related to the STORK.

The MAGUARI.

Ardea-Maguari Gmel.
Maguari Brasilienſibus. Marcg.
Ciconia Americana. Klein. and Briſſ.
The American Stork. Lath.

THE Maguari is a large bird which inhabits the hot parts of America, and was firſt deſcribed by Marcgrave. It is of the bulk of the Stork, and, like it, clatters with its bill, which is long and ſtraight, greeniſh at the root, bluiſh at the point, and nine inches in length. All the body, the head, the neck, and the tail, are inveſted with white feathers, which below the neck are of a conſiderable length and pendulous; the quills and the great coverts of the wing are black gloſſed with green, and when it is cloſed, the quills next the body appear equal to the exterior ones, a property common to all the Marſh birds; the orbits are naked and covered with a bright red ſkin; the

the throat alſo is ſheathed with a ſkin, which may be inflated and formed into a bag; the eye is ſmall and ſparkling, the iris of a ſilvery white; the naked part of the thighs and legs, red: the nails of the ſame colour, broad and flat. We know not whether this bird migrates like the Stork, which it repreſents in the New World. The nature of the climate would ſeem to render the change of reſidence unneceſſary both to the Maguari, and all the other birds of thoſe countries, where the ſeaſons are conſtantly uniform, and the earth, teeming with unceaſing fertility, preſents them a perpetual repaſt. We are ignorant, too, of all the other habitudes of this bird, and indeed of almoſt all the facts relating to the natural hiſtory of thoſe vaſt regions of America.

But can we complain of this neglect, or even wonder at it, when we reflect on the character of the people, whom Europe has, for ages paſt, ſent into the new climates; men, whoſe eyes are ſhut to the beauties of nature, and whoſe hearts are ſtill more impenetrable to the ſentiments which the contemplation of it inſpires.

The COURICACA.

Tantulus Loculator. Linn. Gmel. and Klein.
Curicaca Brasilienſibus. Macrg.
Numenius Americanus Major. Briſſ.
The Wood Pelican. Cateſby.
The Wood Ibis. Penn. and Lath.

THIS bird is a native of Guiana, of Braſil, and of ſome countries of North America, which it viſits. It is as large as the Stork, but its body is more ſlender and longer ſhaped; nor would it reach the height of the Stork, but for the length of its neck and legs, which are larger in proportion: it differs alſo by the ſhape of its bill, which is ſtraight three fourths of its length, but curved at the point, very ſtrong and thick, not furrowed, and of an even roundneſs, gradually enlarging as it approaches the head, where it is ſix or ſeven inches in girth, and near eight inches long; and this large bill is of a very hard ſubſtance, and ſharp at the edges: the back of the head and the arch of the neck are covered with ſmall brown feathers, ſtiff, though looſe: the quills of the wing and of the tail are black, with ſome bluiſh or reddiſh reflections; all the reſt of the plumage is white:

white: the front is bald, and only covered, like the orbits, by a dull blue ſkin: the throat, which is equally naked of feathers, is inveſted with a ſkin capable of inflation and extenſion; which has induced Cateſby to term it, very improperly, *the Wood Pelican* In fact, the ſmall bag of the Couricaca differs little from that of the Stork, which alſo can dilate the ſkin of its throat; whereas the Pelican carries a large ſac under its bill, and, beſides has its feet palmated. Briſſon has committed an overſight in referring the Couricaca to the genus of Curlews, to which it bears no ſort of reſemblance. Piſo appears to be the cauſe of this error, by the compariſon which he draws between this bird and the *Indian Curlew of Cluſius*, which is the Red Curlew, (Scarlet Ibis, *Lath.*) and this miſtake is the leſs pardonable, as in the preceding line, Piſo had repreſented it as equal in bulk to the Swan. He had better reaſon to compare its bill to that of the Ibis, which differs in fact from the bill of the Curlews.

This large bird is, according to Marcgrave, frequent on the river of Seregippa or of St. François. It was ſent to us from Guiana, and it is the ſame with what Barrere denominates *the Curved Bill Crane*, *and*

and the Great American Curlew *; an appellation which might have deceived thoſe who reckon this bird a Curlew, but which Briſſon, by another miſtake, refers to the Jabiru.

Cateſby tells us, that every year, numerous flocks of Couricacas arrive in Carolina about the end of ſummer, which is the rainy ſeaſon in that country. They haunt the Savannas, which are now overflowed; they ſit in great numbers on the tall cypreſſes. Their altitude is very erect, and their ponderous bill is ſupported by reſting it on their neck reclined. They retire before the month of November. Cateſby adds, that they are ſtupid birds, cannot be ſcared, and are eaſily ſhot; and that their fleſh is excellent, though they feed on fiſh, and aquatic animals. [A]

* Of this number is Klein; and to expreſs the ſac under the throat of this bird, he frames the fictitious and barbarous appellation of *Tantulus Loculator* (from *loculus*); and miſled by the falſe name of *Pelican*, he refers to Chardin, and applies to it the Perſian names of *Tacab* and *Miſe*, which probably belong to the Pelican, but which ſurely belong not to a bird of Guiana.

[A] Specific character of the *Tantalus Loculator*. "Its face is bluiſh; its bill reddiſh; its legs, the quills of its wings and of its tail, are black; its body white."

The J A B I R U.

Myčteria Americana. Linn. and Gmel.
Jabiru Braſilienſibus. Marcgr.
Ciconia Braſilienſis. Briſſ.
The American Jabiru. Lath.

As nature has multiplied the reptiles on the low grounds overflowed by the waters of the Amazon and of the Oronooco, ſhe has alſo created birds to deſtroy theſe pernicious creatures: ſhe ſeems even to have proportioned their ſtrength to that of the enormous ſerpents which they are deſtined to combat, and their ſtature to the depth of the mud where they are appointed to wade. One of theſe birds is the Jabiru, which is much larger than the Stork, taller than the Crane and twice as thick; and, if force and magnitude confer precedence, it may be ranked the firſt of the inhabitants of the marſh.

The bill of the Jabiru is a powerful weapon; it is thirteen inches long, and three inches broad at the baſe; it is ſharp, and flat, and edged at the ſides, like a hatchet, fixed into a large head, and ſupported by a thick and nervous neck; this bill conſiſts of hard horn,

N.° 174

THE AMERICAN JABIRU.

horn, and is ſlightly bent upwards; a character of which the firſt trace may be perceived in the bill of the Black Stork: the head, and two thirds of the neck of the Jabiru, are covered with a black naked ſkin, ſhaded on the occiput with ſeveral gray hairs; the ſkin under the neck, for a length of four or five inches, is of a vivid red, and forms a broad and beautiful collar, the plumage of the bird being entirely white; the bill is black; the thighs ſtout, covered with large ſcales, black like the bill, and featherleſs for the ſpace of five inches, the leg is thirteen inches; a membranous ligament appears on the toes, and connects more than an inch and half of the outer toe to the inner one.

Willughby ſays, that the Jabiru is at leaſt as large as the Swan; which is true, conſidering, however, that the Swan is of a longer and more ſlender ſhape, and that the Jabiru has exceeding tall legs. He adds, that its neck is as thick as a man's arm; which is likewiſe true. But that author mentions, that the ſkin on the back of the neck is white, and not red; which may be owing to the difference between the dead and the living ſubject. The tail is broad, and extends not beyond the cloſed wings; the bird, when ſtanding, is at leaſt four feet

feet and an half in perpendicular height, and if extended, it would, considering the length of the bill, measure near six feet. It is the largest bird in Guiana.

Johnston and Willughby have only copied Marcgrave on the subject of the Jabiru; they have even copied his figures with all their faults: and there is in Marcgrave a confusion, or rather a mistake of the editor, which our nomenclators, far from correcting, have really increased. We shall endeavour to clear up the matter.

"The Jabiru of the Brasilians, which the Dutch term *negro*, says Marcgrave, has a thicker and even a longer body than the Swan; the neck is as thick as a man's arm; the head proportionally large; the eye black; the bill black, straight, twelve inches long, two and an half broad, sharp at the edges; the upper mandible a little raised, and stronger than the lower; all the bill is slightly curved upwards."

These characters are alone sufficient to mark out the Jabiru of Guiana, which we have just described from nature. But we are surprized to find in Marcgrave, under the foregoing account, the figure of a bird with a slender body, and a bill arched downwards; in short, except the thickness of its neck, differing in every respect from

from his deſcription. On caſting our eyes however on the other page, we perceive under his *Jabiru Guacu Petiguarenſibus*, or *Nhandu Apoa Tupinambis*, which, he ſays, *is of the bulk of the Stork, and its bill arched downwards*, a large bird in an erect poſture, with a thick body, and with a bill arched upwards, and which we readily diſcover to be the Great Jabiru, the real ſubject of the preceding deſcription, except only the thickneſs of the neck, which is not repreſented in the figure. Here therefore is a double miſtake, one in the engraving, and another in the tranſpoſition; ſo that the thick neck of the Jabiru has been given to the *Nhandu Apoa*, which has been placed below the deſcription of the Jabiru, while the figure of this bird has been placed below the deſcription of the *Nhandu Apoa*.

All that Marcgrave ſubjoins ſerves to point out this miſtake, and prove the juſtneſs of our remark. The Jabiru, he tells us, has ſtrong legs, black and ſcaly, and two feet high; all the body is covered with white feathers; the neck is naked, two-thirds covered with black ſkin from the head, and forming below a circle, which he aſſerts to be white, but which we believe is red in the living bird:—This is preciſely the character, in all its features, of our Great Jabiru of Guiana *. Piso has fallen into no such mistake as Marcgrave; he gives the true figure of the Great Jabiru under its true name of *Jabiru Guacu*, and he says that it occurs at the sides of lakes and rivers in remote places; that its flesh, though commonly very dry, is palatable. This bird grows fat in the rainy season, and then the Indians feast on it; they easily kill it with a fowling piece, or even with their arrows. Piso also remarks in the quills of the wings a red reflection, which we could not perceive in the bird sent us from Cayenne, but which may appear in the Jabiru of Brazil.

* Dr. Grew describes a head of the Jabiru, which is exactly like that of the Cayenne Jabiru. The great bill of this bird occurs in most cabinets, as that of an unknown species.

The NANDAPOA.

Myćteria Americana. Linn. and Gmel.

THIS bird, which is much ſmaller than the Jabiru, has however been termed the Great Jabiru (*Jabiru Guacu*) in ſome countries, where the true Jabiru was probably ſtill unknown: but its real Braſilian name is *Nandapoa*. Like the Jabiru, it is featherleſs on the head, and on the top of the neck, and covered only by a ſcaly ſkin: It is diſtinguiſhed from that bird by its bill being *arched downward*, and only ſeven inches long. This bird is nearly of the bulk of the Stork; the crown of the head is covered by a bony protuberance of grayiſh white; the eyes are black; the ears are large and wide; the neck is ten inches long; the thighs are eight, and the legs ſix, and they are of an aſh colour; the quills of the wing and of the tail which does project beyond the wing, are black, with a reflection of a fine red in thoſe of the wing; the reſt of the plumage is white; the feathers below the neck are ſomewhat long and pendant. The fleſh of this bird has a pleaſant taſte, and is eaten after it is ſkinned.

It is evident that this ſecond deſcription of Marcgrave ſuits his firſt figure, in the ſame manner as his ſecond one correſponds to the deſcription of the Braſilian Jabiru, or our Great Jabiru of Guiana, which is undoubtedly the ſame bird. Such in Natural Hiſtory is the confuſion occaſioned by a ſlight miſtake, which goes on increaſing, when nomenclators implicitly copy each other, and multiply books, to the great detriment of ſcience.

No. 175.

THE CRANE.

The CRANE.

La Grue. Buff.
Ardea-Grus. Linn. and Gmel.
Grus *. All the Naturalifts.

OF all the migratory birds, it is the Crane which undertakes and performs the boldeft and moft diftant journies. Originally a native of the North, it vifits all the temperate climates, and even the regions of the South. It is feen in Sweden †, in the Orknies of Scotland ‖, in Podolia ‡, in Volhinia §, in Lithuania, and in the whole of the North of Europe ¶. In autumn, it alights in our low fens and our

* It is remarkable, that in all languages the name of the Crane imitates its cry: In Greek Γεϱανος. In Latin *Grus*: In Italian *Gru*, or *Grua*: In Spanifh *Grulla*, or *Gruz*: In German *Krane*, or *Kranich*: In Swedifh *Trana*: In Danifh *Trane*: In Swifs *Krye*: In Polifh *Zoraw*. It is uncertain whether the Crane had a name in Hebrew, at leaft it cannot be determined in that barren but obfcure language. In Jeremiah, chap. viii. *Agur* is by Bochart thought to be the *Crane*; but the vulgate renders it *Stork*; and again in Ifaiah xxxviii, the fame word is interpreted *Swallow*.

† *Fauna Suecica.*

‖ Sibbald, *Scotia Illuftrata.*

‡ Rzaczynfki,

§ Klein.

¶ Belon.

ſown fields; then it haſtens to the South, from whence it returns with the ſpring, and again penetrates into the Northern countries, thus completing its circuit with the round of the ſeaſons.

Struck with theſe continual migrations, the ancients termed it the bird of Libya *, or the bird of Scythia †; ſince, by turns, they ſaw it arrive from both of theſe oppoſite extremities of the then known world. Herodotus and Ariſtotle make Scythia to be the ſummer abode of the Cranes: and thoſe which halt in Greece really deſcend from that extenſive region. Theſſaly is called by Plato *the Paſture of the Cranes*; there they alight in flocks, and covered alſo the Cyclades. Heſiod marks the time of their paſſage, when he ſings, "that the huſbandman ſhould obſerve the ſcream of the Crane from aloft in the clouds, as the ſignal to begin his ploughing ‖." India and Æthiopia were the countries aſſigned as its ſouthern reſidence §.

* Euripides.

† Ariſtotle.

‖ Φραζεσθαι δ' ευτ' αν φωνην γερανου επακουσης
Υψοθεν εκ νεφεων ενιαυσια κεκληγυιης
Ητ' αροτοιο τε σημα φερει, ———

Heſiodi Opera et Dies. Lib. ii. 66.

§ Higher Egypt is full of Cranes during the winter; they arrive from the northern countries, to ſpend the cold months." *Voyage de Granger, p.* 238.

Strabo

Strabo ſays, that the people of India eat the eggs of Cranes †; Herodotus, that the Egyptians cover bucklers with their ſkins ‡: And to the ource of the Nile the ancients referred the ſcene of their combats with the Pygmies, *a race of little men*, ſays Ariſtotle, *mounted on ſmall horſes, and who live in caves* §. Pliny places the country of the Pygmies among the remoteſt mountains of India, beyond the fountains of the Ganges; he relates, that their climate was ſalubrious, perpetually mild, and fanned by the northern breeze. "It is reported, he continues, that ſitting on the backs of rams and of goats, and armed with bows, the whole nation deſcends in the ſpring, and conſumes the eggs and young of theſe birds; and that this expedition laſts during the ſpace of three months, otherwiſe it could not reſiſt the invaſions of future flocks ||." In another part of his work he tells us, that the northern part of Thrace was poſſeſſed by a tribe of Scythian's, and he adds, that the towns Aphrodiſias, Libiſtos, Zigere, Borcome, Eumenia, Parthenopolis, Gerania, are ſaid to have been inhabited by

† Lib. xv.
‡ Lib. vii.
§ *Hiſt. Anim.* Lib. viii. 15.
|| Lib. vii. 2.

the race of Pygmies, whom the Barbarians call Catizi, and believe to have been destroyed by the Cranes ¶.

These ancient fables * are absurd, it will be said; and I will allow it: but popular traditions generally contain important facts, though obscured by exaggeration, or concealed under the veil of allegory. I am therefore strongly disposed to believe, that this story alludes to some singularities in the history of the Crane. It is well known that the apes, which rove in large bodies in most parts of Africa and India, wage continual war with the birds; they seek to surprize them in the nest, and lay perpetual snares for them. The Cranes, on their arrival, find these enemies assembled, perhaps in numbers to attack, with more advantage, their new and rich prey. The Cranes, confident in their own strength, enured to fight by their disputes with each other, and naturally prone to combat †, as their attitudes, their movements, and the order with which they marshal, sufficiently evince,

¶ Lib. iv. 11.

* They are anterior to the time of Homer, who compares (*Iliad III.*) the Trojans to the Cranes fighting clamourously with the Pygmies.

† "The Cranes fight so obstinately with each other, that they may be caught while engaged." Aristotle, *Hist. Anim.* Lib. ix. 12.

make

make a vigorous defence. But the apes, obſtinately bent on plundering the eggs and the young, return repeatedly in troops to renew the battle; and as, by their ſubtlety, their gait and poſture, they imitate human actions, they appeared a band of little men to the rude ſpectators, who viewed them from a diſtance, or who, captivated by the marvellous, choſe to embelliſh their relations ‡—Such is the origin and hiſtory of theſe fables.

The Cranes fly very lofty, and arrange themſelves for their expedition; they form a triangle, almoſt iſoſceles, the better to cleave the air. When the wind freſhens, and threatens to break their ranks, they collect their force into a circle; and they adopt the ſame diſpoſition when the Eagle attacks them. Their voyage is ofteneſt performed in the night; but their loud ſcreams

‡ This is not the firſt time that troops of apes have been taken for hordes of Barbarians; not to mention the battle which the Carthaginians fought with the Orang-Outangs on the coaſt of Africa, and the ſkins of three females hung up in the temple of Juno at Carthage as the ſkins of three wild women (*Hanno's Periplus*, Hague 1677, p. 77.) Alexander, in his march through India, would have fallen into this error, and have ſent his phalanx againſt an army of Pongos, if king Taxilus had not undeceived him, by remarking to him, that this multitude which he ſaw following on the heights conſiſted of peaceful animals, attracted by curioſity; but, in fact, far leſs ſenſeleſs and leſs bloody than the plunderers of Aſia. See Strabo, *Lib.* xv.

betray

betray their courſe. During this nocturnal paſſage, the leader frequently calls to rally his forces, and point out the track; and the cry is repeated by the flock, each anſwering, to give notice that it follows and keeps its rank.

The flight of the Crane is always ſupported uniformly, though it is marked by different inflections; and theſe variations have been obſerved to indicate the change of weather: a ſagacity that may well be allowed to a bird, which, by the vaſt height to which it ſoars, is able to perceive or to feel the diſtant alterations and motions in the atmoſphere *. The cries of the Cranes during the day forebode rain; and noiſy tumultuary ſcreams announce a ſtorm: if in the morning or evening they riſe upwards, and fly peacefully in a body, it is a ſign of fine weather; but if they keep low, or alight on the ground, it menaces a tempeſt †. Like all other large birds, except the rapacious tribe, the Crane has much difficulty in commencing its flight. It runs a few ſteps, opens its wings, mounts a little way, and then, having clear ſpace, it diſplays its vigorous and rapid pinions.

* Ariſtotle, Lib. ix. 10.
† *Id. ibid.*

When

When the Cranes are assembled on the ground, they set guards during the night; and the circumspection of these birds has been consecrated in the hieroglyphics, as the symbol of vigilance. The flock sleep with their head concealed under their wing, but the leader watches with his head erect, and if any thing alarms him, he gives notice by a cry *. It is to direct their retreat, says Pliny, that this leader is chosen †. But without supposing an authority conferred, as in human societies, we must allow that these animals are prompted by social instinct to congregate, and to follow the one which calls, which precedes, or which regulates their course. Accordingly, Aristotle ranks the Crane at the head of the gregarious birds ‡.

The first cold days of autumn inform the Cranes of the revolution of the season; and then they retire to milder climates. Those of the Danube and of Germany pass into Italy §. They appear in France in the

* *Id. ibid.* Also Pliny, *Lib.* x. 30.

† *Lib. x.* 30.

‡ Aristotle, *Hist. Anim.* Lib. viii. 12. Festus gives the etymology of the word *congruere, quasi ut grues convenire.*

§ Willughby says that they are pretty common in the markets at Rome; and *Rzaczynski* affirms that a few remain during the winter in Poland, about certain marshes which never freeze.

months

months of September and October, and even in November, when the latter end of the autumn is soft and temperate; but most of them push rapidly on their journey and never halt: they return early in the spring, in March or April. Some of them stray from the main body, or hasten back; for Redi saw them on the 20th of February near Pisa. It would appear that formerly they spent the whole winter * in England; since in Ray's time about the beginning of the present century, they frequented, in great flocks, the fens of Lincolnshire and Cambridgeshire: but the authors of the British Zoology inform us that now they very seldom visit the island of Great Britain, where however they have been known to breed; that there was a penalty against breaking their eggs; and that according to Turner the young Cranes were common †.

* In the text it is put *été* or *summer*; but the passage quoted below shews that this is a misprint, or an inadvertancy.

† "This species we place among the British birds on the authority of Mr. Ray, who informs us, that in his time they were found during the winter in large flocks in Lincolnshire and Cambridgeshire: but on the strictest enquiry we learn, that at present the inhabitants of those counties are entirely unacquainted with them. Though this species seems to have forsaken these islands at present, yet it was formerly a native, as we find, in Willughby, that there was a penalty of twenty-pence for destroying an egg of this bird; and

But I know not what degree of credit is due to the aſſertion of theſe Zoologiſts: no reaſon appears why the Cranes ſhould have forſaken England, nor have theſe authors told us whether the fens in the counties of Cambridge and Lincoln have been drained. There is certainly no diminution of the ſpecies, for we learn from Linnæus that in Sweden it is as numerous as ever: it is indeed in the Northern countries among the marſhes that the Cranes generally breed ‡. On the other hand, Strabo § aſſures us that they neſtle in India; which would prove that like the Storks, they have two hatches annually in two oppoſite climates. The Cranes lay only two eggs ||; and the young ones are ſcarcely reared when the ſeaſon of their departure arrives, and they muſt employ their newly acquired ſtrength in accompanying their parents in their rout *.

and Turner relates that he has very often ſeen their young in our marſhes." Britiſh Zoology.

N. B. The laſt word *marſhes* is, by a ridiculous miſtake, tranſlated *marchés* or *markets*, and M. Buffon immediately ſubjoins that "the fleſh of the Crane is delicate, and was much eſteemed by the Romans."

‡ Klein. Rzaczynſki. Belon.

§ Geograph. *Lib. xv.*

|| Ariſtotle, *Hiſt. Anim.* Lib. ix. 18.

* Belon.

The

The Crane is caught in its paſſage with the nooſe †; and the Eagle and Falcon are alſo flown at it ‡. In certain diſtricts of Poland, the Cranes are ſo numerous that the peaſants are obliged to build huts in the midſt of their fields of buck-wheat, to drive them off §. In Perſia, where they are likewiſe very common ||, it is the prerogative of the prince to hunt them *. The ſame is the caſe in Japan; and that privilege, joined to ſuperſtitious motives, has induced the people to treat the Cranes with

† *Tum gruibus pedicas, & retia ponere cervis.* Virgil *Georg. I.*

‡ Bernier ſaw in the Mogul's dominions the chaſe of the Crane "This chaſe is ſomewhat amuſing; it is pleaſant to ſee them exerting all their force to defend themſelves in the air againſt the birds of prey. They kill theſe ſometimes, but as they want dexterity in turning themſelves, ſeveral good birds will in the iſſue prove victorious." *Hiſt. Gen. des Voyages, tom. x. p.* 102.

§ Rzaczynſki.

|| Lettres Edifiantes, *twenty-eighth Collection.* P. 317.

* "At early dawn, the king (of Perſia) ſent to inform the ambaſſadors, that he would go with a very few attendants to the chaſe of the Cranes, entreating them not to bring their interpreters, that the Cranes might not be ſcared by a multitude, nor the pleaſure of the ſport diſturbed by noiſe. It began with the day. A covered way had been made under ground, at the end of which was the plain, where corn had been ſcattered; the Cranes came in great numbers, and more than four-ſcore were caught. The king took ſome feathers to put into his turban, and gave two to each of the ambaſſadors, who ſtuck them into their hats." *Voyage d' Olearius, Paris,* 1656, *tom. I. p.* 309.

great

great reſpect †. They have been reared and trained in the domeſtic ſtate; and as they naturally ſport in various capers, and then walk with oſtentatious gravity ‡, they can be inſtructed to perform dances §.

We have ſaid, that birds, having a ſofter texture of bones than quadrupeds, live proportionally longer; the Crane affords an example. Many authors mention its longevity: the Crane of the philoſopher *Leonicus Tomæus*, in Paulus Jovius, is famous; he fed it forty years, and it is ſaid that they died together.

Though the Crane is granivorous, as the conformation of its ſtomach ſeems to indicate, and as it generally arrives after the grounds are ſowed to gather the ſeeds which

† The wild birds are become ſo familiar in the iſlands of Japan, that many ſpecies may even be ranked with the domeſtic animals; the principal is the *Tſuri*, or the Crane, which a ſpecial law has reſerved for the diverſion or uſe of the emperor. This bird and the tortoiſe are held to be animals of good omen; an opinion founded on the long life which is aſcribed to them, and on a thouſand fabulous anecdotes with which their hiſtory is filled. The apartments of the emperor, and the walls of the temples are decorated with their figures, as for the ſame reaſon we there ſee thoſe of the fir and of the bamboo. Never do the people call a Crane by any other name than *O tſuriſama*, that is, *My Lord Crane*." Kœmpfer, *Hiſt. Nat. du Japon*, tom. I. p. 112.

‡ Klein.

§ Pliny, *Lib.* x. 39.

the

the harrow has not covered ‖; yet it prefers inſects, worms and ſmall reptiles: and, for this reaſon, it haunts the fens, where it obtains its chief ſubſiſtence.

The membrane, which in the Stork connects the three toes, joins only two in the Crane, the middle and outer ones. The *trachea arteria* is of a very remarkable ſtructure; for, perforating the *ſternum*, it deſcends to a conſiderable depth, and, after making ſeveral twiſtings, it returns by the ſame aperture, and paſſes to the lungs. To the circumvolutions of that organ and to the echo which they produce, we muſt attribute the ſtrong voice of this bird *. The ſtomach is muſcular, and there is a double *cæcum* †; and in this reſpect the Crane differs in its internal ſtructure from the Herons, which have only one *cæcum*. It is diſtinguiſhed externally by its magnitude, by its ſhort bill, by its fuller habit, and by the colour of its plumage. Its wings are very large, and furniſhed with ſtrong muſcles ‡, and contain twenty four quills.

The

‖ Aldrovandus conjectures that the Greek name of the Crane, γερας is compounded of γη, *the earth* and ερευνω, to *ſearch*; becauſe it gathers ſeeds on the ground.

* Belon, *Nat. des Oiſeaux*, p. 187.— Duverney, *Hiſt. de l' Acad des Scien. Ann.* 1666.—1686. *tome. ii. p.* 6.

† Willughby.

‡ The prodigious force of muſcles requiſite to urge ſuch a diſtant flight, had probably given riſe to the prejudice which

The Crane has an erect gait, and a slender figure; the whole field of its plumage is of a fine waved light ash-colour, except the tips of the wings and the covering of the head; the great quills of the wing are black; those next the body extend beyond the tail; the middle and great coverts are of a pretty light ash-colour on the outside, and black on the inside, as well as at the point; from under the last and nearest the body, rise broad unwebbed feathers, which swell into a tuft, and fall back gracefully, and by their flexibility, their position, and their texture, resemble the plumes of the Ostrich; the bill measures four inches from the tip to the corners; it is straight, pointed, compressed at the sides §; its colour is of a greenish black which whitens at the tip: the tongue is broad and short, hard and horny at the end: before the eyes, and on the front and the skull, there are black hairs so thinly scattered that those parts appear almost bare: that skin is red in the living animal; a difference which Belon marks between the male and female,

which prevailed in the time of Pliny, that a person could support any fatigue, who wore a Crane's sinew. *Lib. xxviii.* 87.

§ It has given name to the plant *Geranium* or *Crane's-bill*, which, in all its species, bears this character of fructification.

 in

in which it is not red: ſome feathers of a very deep aſh colour cover the back of the head, and extend a little on the neck: the temples are white, and this colour deſcends three or four inches from the top of the neck: the cheeks, from the bill and below the eyes, and alſo the throat and a part of the foreſide of the neck, are of a blackiſh aſh-colour.

White Cranes ſometimes occur; Longolius and others ſay that they have ſeen them. Theſe are only varieties of the ſpecies, which admits alſo of very conſiderable differences in regard to bulk. Briſſon ſtates the length of his Crane, from the point of the bill to the end of the tail, to be three feet and an inch, and meaſured to the toes, three feet nine inches: he deſcribes therefore a very ſmall bird ||. Willughby makes his to be five Engliſh feet, and ſays that it weighs ten pounds; in which, Ornithologiſts agree with him. In the King's cabinet, there is a Crane, reckoned indeed among the largeſt, which is four feet two

|| Rzaczynſki appears to admit two families of Cranes; to the ſmaller one he attributes ſome particular properties, which do not ſeem however to conſtitute a different ſpecies; the leſſer Cranes bear creſts hoary behind the ears, but blackiſh under the throat. The ſmall breed occurs in Volhinia and in the Ukraine; the large one in Cujavia, and both together in Podolia.

inches

inches in perpendicular height, and if ſpread out, would meaſure more than five feet between the tip of the bill and the toes; the naked part of the thighs is four inches, the legs are black, and only ten inches and a half.

Conſidering its force of wing and its migratory inſtinct, we cannot be ſurprized that the Crane ſhould viſit every climate. Yet we ſuſpect that it never advances further South than the tropic: all the countries where the ancients placed the winter reſidence of the Crane, Lybia, the regions ſituate at the origin of the Nile, of the Indus, and of the Ganges, are within this limit, which alſo bounded the Geography of antiquity. The Cranes, natives of the North, ſeek only a moderate degree of temperature, and not the ſcorching heats of the Torrid Zone. The fens which invite them never occur in the midſt of arid plains and burning ſands; and if a few tribes following gradually the chain of mountains, where the heats are mitigated, at laſt paſs the equator and advance into the South, they become loſt in thoſe countries, diſmembered from the reſt of the ſpecies, and never join in the migrations to the North. Such in particular are the Cranes which, Kolben ſays, are numerous at the Cape of Good

Good Hope; and exactly the ſame with thoſe of Europe; a fact which we ſhould ſcarce have admitted on the ſingle teſtimony of that traveller, had not others found Cranes in almoſt as high Southern latitudes, as in New Holland * and in the Philippines where there are two ſpecies †.

The Eaſt India Crane ſuch as the moderns have obſerved it, ſeems not to differ in ſpecies from the European; it is ſmaller, its bill rather longer, the ſkin on the top of its head, red and hard, and extending to the bill; in other reſpects, it is exactly like the ordinary one, and its plumage is of the ſame aſh-gray. This is the deſcription which Willughby gives of one which he ſaw alive in St. James's Park. Edwards deſcribes another Crane brought from India ‡. It was, he tells us, a large fine bird, ſtronger than ordinary, and its extended height near ſix feet. It was fed with barley and other grain; it laid hold of its food with the point of its bill, and giving its head a quick toſs backwards, it threw it into the throat; a red naked ſkin

* Captain Cook's firſt Voyage.

† Camel, No. 285, *Philoſophical Tranſactions.*

‡ Ardea-Antigone. *Linn. and Gmel.*
Grus Orientalis Indica. *Briſſ.*
The Greater Indian Crane. *Edw.*
The Indian Crane. *Lath.*

ſkin with a few black hairs covered the head and the top of the neck; all the plumage was of a blackiſh aſh-colour, only rather lighter on the neck; the thighs and legs were reddiſh. In all theſe features, no very preciſe ſpecific difference can be traced; yet Edwards reckons his *Greater Indian Crane* as quite a diſtinct bird from that of Willughby, and what moſt of all induces him to entertain the opinion is the great difference of ſize: and we might agree with him, did not the European Cranes vary widely in that reſpect §. This Crane is probably the ſame with that which inhabits the Eaſt of Aſia oppoſite to Japan ||, and which migrates in winter into Judea, and even deſcends into China, where theſe birds are numerous *.

§ It ſeems impoſſible to draw any certain concluſion from the following paſſage of Marco Polo. "Near the coaſts of the Cianiganians are five ſorts of Cranes: ſome have black wings like Ravens; others are very white with golden ſpangles on their plumage, as in Peacocks tails; others are like ours, and others ſtill are ſmaller, but have their feathers very long and beautiful intermingled with red and black colour; thoſe of the fifth kind are gray, having red and black eyes, and theſe are very large."

|| The Cranes are ſeen in Siberia among the Jakutes. Innumerable flocks appear in the plain of Mangaſea, on the Jeniſca. *Gmel.*

* "The Cranes are very numerous in China; this bird accommodates itſelf to all climates. It is eaſily tamed, and even learns to dance. Its fleſh is reckoned very good food." *Hiſt. Gen. des Voyages, tom. vi. p.* 487.

To the ſame ſpecies we muſt refer the Japan Crane ſeen at Rome, which Aldrovandus has figured and deſcribed. "It was equal in bulk to our Crane, the top of its head was of a bright red, ſtrewed with white ſpots, and the colour of its plumage bordered on white †." Kœmpfer ſpeaks likewiſe of a White Crane at Japan; but as he makes no other diſtinction between it and the gray one, which he mentions in the ſame place ‡, there is every probability that it was only the variety obſerved in Europe. [A]

† Grus Japonenſis Alia. *Aldrov. Johnſt. Charleton, and Klein.*

Grus Japonenſis. *Briſſ.*

‡ "There are two ſorts of Cranes in Japan, the one is white as alabaſter, the other gray or aſh-coloured." *Nat. Hiſt. du Japan*, tom. I. p. 112.

[A] Specific character of the Common Crane, *Ardea-Grus*: "Its head is naked and papillous; the tuft and the quills of the wing, black; its body cinereous; its inmoſt tail-quills, unwebbed." It breeds in fens, and lays two bluiſh eggs. It reſts on one foot. Linnæus aſſerts, that, in its paſſage, it flies at the height of three miles.

The COLLARED CRANE.

Ardea-Antigone. Var. Gmel.

THIS Crane ſeems to differ ſo much from the common ſpecies that it could not be claſſed with it by the ſame analogies as the preceding varieties Beſides that it is much ſmaller than the ordinary Crane, its head proportionably thicker, and its bill larger and ſtronger; it has the top of its neck decorated by a fine red collar, with a broad white belt under it, and all the head naked, and of an uniform reddiſh gray, without thoſe ſtreaks of white and black, which encircle the head of the common Crane: the tuft of the tail alſo is of the ſame bluiſh gray with the reſt of the body.

CRANES of the New Continent.

The WHITE CRANE.

Ardea Americana.

The Hooping Crane. Catefby. Edw. Penn. and Lath.

IT is extremely probable that the Cranes have migrated from the one continent into the other, fince they haunt the Northern parts of Europe and of Afia. Accordingly, we find in America a White Crane, and one or two kinds of Gray or Brown Cranes. But the White Crane, which in our Continent is only an accidental variety, appears to have formed in the other a permanent race, difcriminated by ftrong and decided characters; fo that we may conceive it as anciently feparated from the common fpecies, and long modified by the influence of climate. It is as tall as our largeft Cranes, but of a ftronger and thicker make, its bill longer, its head bigger; its neck and legs not fo flender: all the plumage is white, except the great quills of the wings, which are black, and the head, which

which is brown; the crown is callous and covered with black hairs, ſtraggling and delicate, under which the reddiſh ſkin appears naked; a ſimilar ſkin covers the cheeks: the tuft of looſe feathers in the tail is flat and pendant: the bill is furrowed above, and indented at the edges near the tip; it is brown, and ſix inches long. Cateſby has deſcribed this bird from an entire ſkin given him by an Indian, who told him that theſe birds frequent, in great numbers, the lower parts of the rivers near the ſea in the beginning of ſpring, and return to the mountains in ſummer. "This fact, ſays Cateſby, has been ſince confirmed by a white, who informed me that theſe Cranes are very noiſy, and are ſeen in the Savannas at the mouth of the Altamaha, and other rivers near St. Auguſtine in Florida, and alſo in Carolina; but that they are never found further North."

Yet it is certain that they advance into the higher latitudes: for the ſame White Cranes are found in Virginia *, in Canada †,

* *De Laet. p.* 83. The firſt voyagers to America ſpeak of Cranes which they ſaw: Peter Martyr ſays, that, in the Savannas of Cuba, the Spaniards met with flocks of Cranes twice as large as ours.

† "We have (in Canada) Cranes of two colours; ſome are entirely white, others of a gray gridelin: they all make excellent ſoup." *Charlevoix. Hiſt. de la Nov. France tom. iii. p.* 135.

and

and even in Hudſon's Bay, as Edwards remarks. [A]

[A] Specific character of the Hooping Crane, *Ardea Americana:* " Its top, its nape, and its temples, are naked and papillous; its front, its nape, and its primary wing quills are black; its body is white." The extreme length is five feet ſeven inches. We ſhall extract the following paſſage relating to theſe birds, from Mr. Pennant's Arctic Zoology. " They make a remarkable hooping noiſe: this makes me imagine theſe to have been the birds, whoſe clamor Captain Philip Amidas (the firſt Engliſhman who ever ſet foot on North America) ſo graphically deſcribes, on his landing on the iſle of Wokokou, off the coaſt of North Carolina "When, ſays he, ſuch a flock of Cranes (the moſt part white) aroſe under us with ſuch a cry, redoubled by many echoes, as if an army of men had ſhouted together." This was in the month of July; which proves, that in thoſe early days this ſpecies bred in the then deſert parts of the ſouthern provinces, till driven away by population, as was the caſe with the common Crane in England; which abounded in our undrained fens, till cultivation forced them entirely to quit our kingdom. *Vol. ii. p.* 442.

The BROWN CRANE.

Ardea Canadensis. Linn. and Gmel.
Grus Freti Hudsonis. Briss.
The Brown and Ash-coloured Crane. Edw.

It is about a third smaller than the preceding: the great quills of its wings are black; their coverts, and the scapular feathers reaching to the neck, are rusty brown, and so are the large flowing feathers disposed near the body: the rest of the plumage is cinereous: the red skin of the head covers only the front and the crown. These differences and that of the size, which in this family of birds varies much, are not perhaps sufficient to separate this species, from that of our Crane, they are at least two kindred species, especially, as their habits and climates are similar; for they usually advance into the North, as far even as Hudson's Bay, where they breed, but, on the approach of winter, again return towards the South, holding their course, it would appear, through the country of the Illinois *, and of

* "Among the Illinois there are many Cranes. *Lettres Edifiantes*, eleventh collection, p. 310.

the

the Hurons †, and pushing onwards to Mexico ‡, or even further. These American Cranes have the same instinct therefore as those of Europe; they travel also from North to South, and this is probably what the Indian meant, who told Catesby, that they retired from the sea to the mountains. [A]

† "In the season the fields (of the Hurons) are all covered with Cranes, or *tochingo*, which repair to feed on their corn when planted, or when near ripe. They kill these Cranes with their arrows, but not often, because if this large bird has not its wings broken, or is not shot dead, it easily bears away the arrow in its wound, which in time heals up; as our Canadian missionaries have experienced, a Crane being taken at Quebec which had been struck with a Huron's arrow, three hundred leagues distant; the wound on the rump was healed, and the end of the arrow, with its flint, was inclosed. They are caught sometimes with gins." *Voyage au pays des Hurons*, par le P Sagard Theodat, *Paris*, 1652. *p.* 302, 303.

‡ It is easy to recognise this Crane in the *Toquilcoyotl* of Fernandez. "To the Crane a species may be referred, which is of the same size, and has similar habits and instincts; it is called *Toquilcoyotl*, in imitation of its cry: its whole body is brown mixed with black and cinereous; its head is decorated above with a scarlet spot, &c. *Avi. Nov. Hisp.* cap. cxlviii. p. 44. This North American Crane, which migrates from the regions of the South, Brisson has made his eighth species under the name of *Mexican Crane*, and which Willughby, Klein, and Ray, have termed the *Grus Indica*.

[A] Specific character of the Brown Crane, *Ardea Canadensis*: "Its forehead is naked and papillous, its body cinereous, its wings externally brick-coloured." This species advances from Mexico, to the northern parts of the continent, early in the spring. It appears in Hudson's Bay about the month of May, frequents the lakes and pools, and retires again in autumn.

No. 176

THE NUMIDIAN CRANE OR DANCING BIRD.

Foreign Birds which are related to the CRANE.

The NUMIDIAN CRANE.

La Demoiſelle de Numidie. Buff.
Ardea-Virgo. Linn. and Gmel.
Grus Numidica. Briſſ. and Klein.
The Dancing Bird. Pocock.

THE Demoiſelle of Numidia has all the proportions and the ſhape of the Crane, only on a ſmaller ſcale; its port, its garb, are the ſame; and the ſame diſtribution of colours on the plumage, only the gray is purer, and more pearled. Two white tufts of unwebbed and hairy feathers, falling on each ſide of the head, form a ſort of head-dreſs; long, ſoft and ſilky hairs of the fineſt black lie on the crown of the head; ſimilar feathers deſcend from the fore-part of the neck, and hang gracefully below it; between the black quills of the wings, appear bending tufts, which are long and pendant. This beautiful bird has received the name of *Demoiſelle*, or *Miſs*, on on account of its elegant form, its rich garb, and its affected airs: it makes repeated reverences; it walks with a sprightly ostentation, and it often leaps and springs from gaiety, as if it meant to dance.

This bent, which, in a certain degree, was remarked in the Crane, is so striking in this Numidian bird, that for more than two thousand years, during which it has been known, authors have constantly denominated it from its mimic gestures. Aristotle calls it the actor or comedian *; Pliny, the dancer or vaulter †; and Plutarch mentions its frolics and its address ‡. It appears even to imitate the actions which it beholds: Xenophon, in Athenæus, seems persuaded of this; "for, says he, the fowlers rub their eyes before it with water which they pour into basons, and then filling these up with bird lime they retire, and the bird copying their example rubs its eyes and feet." Accordingly Athenæus terms it *The Imitator of Man* §. The Dancing Bird of Numidia seems also to have copied our vanity; it loves to be seen, and to exhibit itself, it seems even to prefer

* Hist. Anim. *Lib. viii.* 12.

† *Lib. x.* 23.

‡ De Solert. Animal.

§ Ανθρωποειδης.

show

ſhow to its food, and follows after a perſon to ſollicit as it were another look.

Theſe are the remarks of the academicians on the Numidian Cranes, of which they had ſeveral in the *Menagerie* at Verſailles. They compared their ſteps, their poſtures and their geſtures to *Gipſey dances*; and Ariſtotle ſeems to allude to their manner of jumping and vaulting, when he ſays that *they are caught when they dance oppoſite one another*,

Though this bird was famous among the ancients, it was little known, and ſeldom ſeen in Greece or Italy; and confined to its own climate, it enjoyed a ſort of fabulous celebrity. Pliny after terming it in one place the *pantomime*, joins it in another paſſage with ‖ the Syrens, the Griffins, and the Pegaſuſes. It was late before the moderns were acquainted with it; they confounded it with the *Scops* and *Otus* of the Greeks and *Aſio* of the Latins, on account of the odd geſtures of that Owl, whoſe ears were ſuppoſed to be repreſented by the long delicate filaments that hang from each ſide of the head of the Numidian Dancing Bird.

The ſix *Demoiſelles* which were kept ſome time in the menagerie, *came from Nu-*

‖ Lib. x. 49.

midia. We can find nothing more precise in naturalist with regard to the countries which it inhabits *. Travellers have met with it in Guinea †, and it appears to be a native of the tropical parts of Africa. Yet it would not be impossible to reconcile it to our climate, to naturalize it in our court-yards, and even to perpetuate the breed. The Numidian Cranes of the Royal *menagerie* have propagated, and the one which died last, at the age of twenty-four, was hatched in it ‡.

The academicians give very minute details of the internal structure of the six birds which they dissected; the *trachea arteria*, which was of a hard and almost bony substance, was inserted, by a double circumvolution, into a deep groove formed in the top of the sternum; below the trachea, they perceived a bony knot, that had the shape of a *larynx*, parted internally into two by a little tongue, as in the goose and in some other birds; the *cerebrum and cerebellum* together weighed only a dram and a half; the tongue was fleshy above and

* *The Demoiselle of Numidia.* Edw.

† Hist. Gen. des Voyages, *tom. iii p.* 307.

‡ This fact was communicated to us by order of the Marshal duke de Mouchy, governor of Versailles and of the king's *menagerie.*

cartilaginous

cartilaginous below; the gizzard was similar to that of a hen, and like all the granivorous birds, it contained bits of gravel. [A]

[A] Specific character of the Numidian Crane, *Ardea Virgo*: "Its eye-brows are white, with tufts behind stretching far back."

The ROYAL BIRD.

Ardea Pavonina. Linn. and Gmel.
Grus Balearica. Aldrov. Will. Johnſt. &c.
Grus Japonica Fuſca. Petiv.
Pavo Marinus. Cluſius.
The Balearic Crane. Sloane and Will.
The Crowned African Crane. Edw.
The Crowned Heron. Lath.

THIS bird owes its appellation of *royal** to a ſort of crown, which a bunch of feathers, or rather of ſpreading briſtles, forms upon its head. It has a noble port, and, when it ſtands erect, it is four feet high. Fine feathers of a leaden-black, with bluiſh reflections, hang along its neck, and are diſplayed on its ſhoulders and its back: the firſt quills of the wing are black, the reſt of a brown rufous, and their coverts, which are broken into threads, divide the dark ground of the mantle by two large white ſpots; a broad membranous ſkin, which is of a fine white on the temple, and of a bright carnation on the cheek, covers

* The Dutch, who trade on the coaſts of Africa, call it *Kroon-Vogel*, or Crown-Bird.

the

No. 177

THE CROWNED CRANE.

the face, and descends under the bill †; a cap of black down, delicate and close as velvet, decorates the front, and its beautiful crest is thick, wide spread, and composed of bushy sprigs of a pink colour, flat, and wreathed into a spiral; each sprig is beset lengthwise with minute filaments, pointed with black, and terminated by a small pencil of the same colour; the iris is pure white; the bill black, as well as the thighs and the legs, which are still taller than those of the Cranes, which this bird resembles much in its conformation. It differs from them, however, in several important characters; it is a native of the hot climates, the Cranes draw their origin from the cold countries; they have a dusky plumage, while the Royal Bird is decorated with the garb of the South, of that burning zone, where every thing is more brilliant, but more fantastic; more animated, but less graceful than in the temperate climates.

† Of the two figures which Edwards has given for the male and female, the one has only an earlet behind the eye, and the other the two pendants under the throat. This character seems to vary; it occurs not in the description of Clusius, in other respects exact; and probably it depends rather on age than on sex, since the academicians did not find it in one of the subjects which they described, though both were females.

 They

They inhabit Africa, especially Gambra, the Gold Coast, Juida, || Fida, and Cape Verd. Travellers relate, that they are frequently seen by the large rivers; they feed on small fish, and also search in the fields for herbs and seeds; they run very swiftly, spreading their wings, and catching the wind: at other times their pace is slow, with measured steps.

This Royal Bird is gentle and pacific; its defence is its stature, and the rapidity with which it runs or flies. It is less afraid of man than of its other enemies; and seems to approach with chearfulness and confidence. We are assured, that at Cape Verd this bird is half domestic, and that it comes into the court-yards to eat grain with the pintadoes and other fowls. It perches in open air to sleep, like the peacock, whose cry it is said to imitate; and, as it also bears a resemblance by the tuft on its head,

|| *Hist. Gen. des Voyages, tom. iv. p.* 355. It appears that the Europeans on these coasts have given the same name of *Royal Bird* to a quite different species. "Smith distinguishes two kinds of *Crown-Birds*; the first has its head and neck green, its body of a fine purple, its wings and tail red, and its tuft black; it is nearly as large as the great parrots. The other kind (and this is the true Royal Bird) is of the shape of a Heron, and is not less than three feet high; it feeds on fish: its colour is a mixture of blue and black, and the tuft, with which it is crowned, resembles more hogs' bristles than feathers." *Id. p.* 247.

head, it has been termed by ſome naturaliſts the *Sea Peacock* ‡; others have called it the *Short-tailed Peacock* §, and others have aſſerted that it is the ſame with the *Balearian Crane (Grus Balearica)* of the ancients, but which is not at all proved ¶: for Pliny, the only one who has mentioned the Balearian Crane, does not ſufficiently characterize it; *both the Woodpecker* (ſays he) *and the Balearian Crane wear a tuft* *; but no two things are leſs ſimilar than the ſmall creſt of the Woodpecker, and the crown of the Royal Bird, which beſides has ſeveral remarkable properties that Pliny might have pitched on. If, however, it was formerly brought to Rôme from the Balearian iſlands, where it is no longer found, it would corroborate my poſition, that animals are making a gradual progreſs from the North to the South.

We received this bird from Guinea, and kept it ſome time in a garden. It pecked the herb, particularly the core of lettuce and ſuccory. But the food which agrees beſt with it, is rice, either dry, or ſlightly

‡ Cluſius. *Exotic. Lib. v.* 2.

§ Johnſton, Barrere, Linnæus.

¶ *See* Memoires pour ſervir a l'Hiſtoire des Animaux, tom. iii. 2.

* *Cirros pico martio et Grui Balearicæ*, Lib. xi. 37.

 boiled,

boiled, previously soaked in water, or at least washed and well picked; for it rejects what is of a bad quality, or soiled with dust It seems also to eat insects, and particularly earth-worms; and we have seen it among new-turned mould, gathering worms, and catching small insects on the leaves. It is fond of bathing, and ought to have a shallow tray, in which the water should be renewed from time to time. As a regale, we may throw into the tray a few small living fish, which it eats with avidity, but rejects those that are dead. Its cry resembles much that of the Crane, and is clangorous, like the sound of a trumpet or horn: it is repeated at short intervals when the bird wants food, or when it seeks a place of lodgment for the evening † This cry is also the expression of inquietude and weariness; for it tires if left too long by itself. It is glad of a visit, and if the company (after satisfying their curiosity,) retire without taking farther notice, it follows or walks by their side, and thus makes several turns: and if it is then amused and detained by any novelty, it hastens to rejoin them. When it is in a quiescent posture, it rests on one leg, its large neck is folded back like the lock of a musket, and its body, shrunk and

† This bird has still another voice, or rather an inward clucking, like that of a sitting hen, but harsher.

as it were tottering on its tall limbs, bears horizontally. But if it is alarmed or disturbed, it extends its neck, raises its head, and assumes a stately air, striving by the boldness of its attitude, as it were, to strike awe: its whole body is then almost erect; it advances gravely, and with measured steps; and the dignity of its demeanor, recommended too by the crown which it wears, justly entitles it to the appellation of *royal*. Its long legs, which aid it much in rising, prove troublesome when it alights; and it therefore spreads its long wings to break the fall; but we were obliged to keep them short, by repeatedly clipping the feathers, to prevent it from flying away, which it often attempted to do. It passed the whole of the winter (1778) at Paris, without appearing to suffer from the rigours of a climate so different from its own. It had chosen for itself a room with a fire to shelter it during the night, and it repaired every evening to the door, sounding for admission.

The first birds of this kind were brought into Europe by the Portuguese, in the fifteenth century, when they discovered the Gold Coast in Africa ‡ Aldrovandus cele-

‡ "It seems that these birds are much prized in Europe, since some gentlemen continually ask us to send some." *Voyage de Guinée*, par Guill. Bosman. *Utrecht*, 1705, *Lettre XV*.

brates their beauty ‖, but Belon ſeems not to have known them, and he is miſtaken in aſſerting that the Balearian Crane of the ancients was the Night Heron ‡. Some authors have named them *Japan Cranes* §, which ſeems to imply that they are found in that iſland, and that the ſpecies extends over the whole zone in Africa and Aſia.— The famous Royal Bird, or *Fum hoam* of the Chineſe, of which ſo many marvellous ſtories are told, collected by the credulous Kircher ¶, is only a creature of the imagination, like the dragon, which they paint along with it on their muſlins and porcelain.

‖ *Avis viſu jucundiſſima.*

‡ "We alſo ſaw at Aleppo a bird reſembing a Crane, but ſmaller, having its eyes edged with red, the tail of the Heron, and a weaker voice than the Crane's. We believe it to be what the ancients called the *Balearian Crane.*" *Obſervations de Belon, p.* 159. What leads us to think that this account refers not to the Royal Bird, is, that Belon makes no mention of the crown, a character ſo diſtinct and ſtriking, that it could not have eſcaped this excellent obſerver.

§ Charleton and Petiver.

¶ *See* la Chine illuſtrée. *Amſterdam*, 1670, *p.* 263.

The CARIAMA.

Palamedea Cristata. Linn. and Gmel.
Cariama. Marcg. Piso. Ray. Will. and Briss.
The Crested Screamer. Lath.

We have seen that nature, moving with an uniform pace, connects all her productions by successive gradations. She has by her transitions, filled up the intervals where we would place our divisions, and pause in the wide survey. No void appears in the vast concatenation of the universe, and the most distant parts are linked together. Our systems prove inconsistent, therefore, when they attempt to assign limits which no where exist: nay, productions that stand most detached in our methods, are really connected with others by the greatest number of relations. Such are the Cariama, the Secretary, and the Kamichy. The two first are akin to the birds of prey; the last is, on the contrary, related to the gallinaceous tribe; but all the three resemble most, in their instincts and habits, the marsh birds.

The Cariama is a beautiful bird, which frequents fens, where it feeds like the Heron,

ron, but is larger *; it has long legs and the lower part of its thigh is naked, as in the Marſh-birds; and its bill is ſhort and hooked, as in the birds of prey.

Its neck is high, and its head lofty; under the root of the bill, which is yellowiſh, there is a tuft ſhaped feather; its whole plumage is like that of the Falcon, gray waved with brown; its eyes are brilliant and gold-coloured, and its eye-lids furniſhed with long black hairs; its legs are yellowiſh, and of the toes, which are all connected at their origin by a portion of a membrane, the mid-one is longer than the two ſide ones, of which the inner is the ſhorteſt; the nails are ſhort and rounded †; the hind toe is placed ſo high, that it will not reach the ground; and the heel is thick and round, like that of the Oſtrich. The cry of this bird reſembles that of a Turkey-cock; this is loud, and betrays it to the fowlers, who prize it on account of its tender and delicate fleſh. If we may credit Piſo, moſt of the birds which haunt

* *Egregia avis ſylveſtris cariama ex aquaticorum genere, udoſiſque locis ob prædam delectatur more ardearum, quas mole corporis longè ſuperat.* Piſon. *Hiſt. Nat. & Medic. Ind.* p. 81.

† *Ungues breviuſculi, lunati. Ibid.*

the

the marshes in these hot parts of America are not inferior in the quality of their flesh, to those which inhabit the mountains ‡. He says also that the settlers have begun to domesticate the Cariama. It appears, therefore, that the Cariama, which occurs only in America, is both with regard to its structure and its dispositions, the representative of the Secretary of the Old Continent, which we now proceed to describe. [A]

‡ *Mansuefacta, æque ac sylvestris, assatur & coquitur.* Idem.

[A] Specific character of the *Palamedea Cristata*: "It is unarmed, its front crested."

The SECRETARY,
Or the MESSENGER.

Falco Serpentarius. Gmel. and Miller.
Sagittarius Vosmaer.
Vultur Serpentarius. Lath. Ind.
The Secretary Vulture. Lath. Syst.

THIS bird, alike conspicuous by its magnitude and by its figure, is not only of a new species, but of a distinct and detatched family; so that it eludes and confounds all our artificial classifications. While its long legs seem to shew that it is destined to haunt the fens, its hooked bill declares its affinity to the birds of prey. It has, so to speak, the head of an Eagle joined to the body of a Stork or Crane. To what class then can we refer it? So free and unrestrained is nature, and so rich and so vast the range of her productions!

The Secretary is as tall as a large Turkey Cock; its colours on the head, the neck, the back, and the coverts of the wings, are of a browner gray than that of the Crane, and become lighter on the fore part of the body; the quills of the wings and of the tail, are stained with black, and the thighs

No. 178

THE SECRETARY.

thighs with black waved with gray; a bundle of long feathers, or rather of ſtiff black quills, hangs behind the nerk; moſt of theſe feathers are ſix inches in length, ſome are ſhorter, and a few are gray; all of them are narrow near the baſe, more fully webbed towards the tip, and inſerted on the top of the neck. The ſubject which we deſcribe was three feet ſix inches high, the tarſus alone near a foot; a little above the knee, there were no feathers; the toes were thick and ſhort armed with hooked nails; the mid-one is almoſt twice as long as the lateral ones, which are connected by a membrane near half their length, and the hind toe is very ſtrong: the neck is thick, the bill is ſtrong and cleft beyond the eyes; the upper part of the bill is equally and boldly hooked, nearly as in the Eagle, and it is pointed and ſharp; the eyes are placed in a ſort of naked ſkin, of an orange colour, which extends beyond the outer corner of the eye, and takes its origin at the root of the bill: there is beſides a ſingular character, which ſtill more ſhows the double nature of this bird; it is an eyebrow formed of one row of black hairs, from ſix to ten lines in length *: this feature,

* This eye-brow is fifteen or ſixteen lines long, the laſhes are ranged very cloſe, widening at the baſe, ſcooped into channels, concave below, and convex above.

ture, with the tuft of feathers on the top of its neck, its head like that of a bird of prey, its legs like thoſe of a Shore-bird, form altogether an extraordinary ambiguous aſſemblage.

The habits of this bird are as mixed as its ſtructure. Though it has the weapons of the rapacious birds, it is exempt from their ferocity; and it never employs its bill, either for defence or for attack. It ſeeks ſafety by flight, and often in the hurry of its eſcape, it takes leaps of eight or ten feet high. It is gentle and ſprightly, and ſoon grows familiar. At the Cape of Good Hope it has even been domeſticated, and is pretty frequent in the houſes of the coloniſts: they find it up the country, a few leagues from the ſhore; they take it young from the neſt, and rear it as much for utility as pleaſure, for it deſtroys Rats, Lizards, Toads and Serpents.

The Viſcount de Querhoent has communicated the following obſervations on this bird. "When the Secretary," ſays that excellent obſerver, "lights on a ſerpent, it firſt tires him out, by ſtriking with its wings; it then lays hold of him by the tail, lifts him up on high, and drops him, which it repeats till the ſerpent is killed. It ſpreads its wings, and is thus often ſeen running

and

and flying at once. It nestles in bushes a few feet from the ground, and lays two eggs, which are white with rusty spots. When disturbed, it makes a hollow croaking. It is neither dangerous, nor mischievous. Its disposition is gentle; and I have seen two birds of this kind live peaceably in a court-yard amidst poultry. They were fed with flesh, and were greedy after guts and entrails, which they held under their feet while devouring them, as they would have done a serpent. They slept every night together, and the head of each was turned to the tail of its companion."

This bird, though a native of Africa, seems to accommodate itself to an European climate; it is kept in some of the menageries of England and of Holland. Vosmaer who fed one in that of the Prince of Orange, has made some remarks on its manner of living. "It tears and swallows greedily the flesh that is offered to it, and does not refuse fish. When it prepares itself to repose and sleep, it rests with its belly and breast upon the ground: a cry which it rarely utters resembles that of the Eagle: its most usual exercise is to walk with long steps from one side to another, and for a considerable time without halting or slack-
ening

ening its pace †: it is of a cheerful, quiet, and timid diſpoſition. If a perſon approaches while it runs hither and thither with a ſupercilious air, it makes a continual cracking; but after it has recovered from its fright, it appears familiar, and even prying. When the painter was employed in deſigning it, the bird drew near him, looked attentively upon his paper, ſtretched out its neck, and erected the feathers of its head, as if it admired its figure. It often came, with its wings raiſed and its head projected, to obſerve curiouſly what was doing. It thus approached me two or three times, when I was ſitting at a table in its hut, in order to deſcribe it. On ſuch occaſions, or when it eagerly gathers ſome ſcraps, and in general when it is moved by curioſity or deſire, it briſtles high the long feathers on the back of its head, which commonly fall irregularly on the top of the neck. It was obſerved to moult in the months of June and February; and Voſmaer ſays that, in ſpite of the cloſeſt attention, they could never detect it drinking;

† This quality has probably conferred on it the name of *Meſſenger*, as the bundle of feathers on the top of the neck has procured that of *Secretary*; though Voſmaer ſuppoſes it derived from the appellation of *ſagitary* applied to it, becauſe it often amuſes itſelf by taking a ſtraw in its bill or its foot and darting it repeatedly into the air.

yet

yet its excrements are liquid and white, like thoſe of the Heron. To eat with eaſe, it ſquats on its heels, and thus half-repoſed, it ſwallows its food. Its ſtrength ſeems to lie chiefly in the leg; if a live hen be preſented, it hits her a violent blow with its ſole, and knocks her down by a ſecond ſtroke. It treats rats in the ſame manner. It watches aſſiduouſly before their holes. In general it prefers living to dead animals, and fleſh to fiſh."

It is not long that this ſingular bird has been known even at the Cape, ſince, neither Kolben, nor the others who have deſcribed the productions of that country, make any mention of it. Sonnerat found it in the Philippines, after having ſeen it at the Cape. We perceive ſome differences between his account and the preceding ones, that ſeem to deſerve notice: Sonnerat, for inſtance, repreſents the feathers of the creſt as riſing on the neck at unequal intervals, and the longeſt as placed the loweſt; but neither this order nor proportion occur in the ſubject which I inſpected, theſe feathers being collected into an irregular bunch: he adds, that they are bent in the middle next the body, and that their webs are criſped; Voſmaer gives the ſame ſtatement, yet in the one which we have deſcribed they were ſmooth.

ſmooth. Are theſe differences to be attributed to the objects themſelves, or only to the deſcriptions? But a ſtill greater diſagreement occurs in the colour of the plumage, which Voſmaer ſays is of a bluiſh lead gray, though we found it to be verging on brown: he ſays that the bill is bluiſh, and we found it to be black above and whi e below. The ſubject which we deſcribe is lodged in the cabinet of Dr. Mauduit; the two feathers do not as uſual exceed the tail, they only project five inches beyond the wings when cloſed. But another ſubject, from which our figure was taken, has three long feathers, ſuch as they are deſcribed by Voſmaer and Sonnerat. We conceive this to be the character of the male. Sonnerat is miſtaken when he reckons the bill of the Secretary as *gallinaceous*; which is the more ſtrange, as that naturaliſt remarks that the bird itſelf is carnivorous.

Reflecting on the ſocial and familiar diſpoſitions of this bird, and its facility in domeſtication, we are diſpoſed to think that it would be adviſable to multiply the ſpecies, particularly in our colonies, where it might be ſerviceable in deſtroying the pernicious reptiles and rats.

[A] Specific character of the *Falco Serpentarius*: "It is black, its head creſted, the tip of its tail-quills white, the two middle ones very long."

N°. 178

THE HORNED SCREAMER.

The KAMICHI.

Palamedea Cornuta. Linn. and Gmel.
Anhima Brasilienfibus. Marcg. and Pifo.
Anhima. Briff.
Cahuitahu. Condamine.
The Horned Screamer. Lath.

NATURE is not to be ftudied in the cultivated fields, that fmile under the forming hand of induftry. We muft vifit the burning fands of the tropical regions, and the eternal ice of the pole; we muft defcend from the fummits of mountains into the bed of the ocean; and we muft compare remote wilds and deferts: and fuch magnificent contrafts confer additional fublimity on the fcenes of the univerfe. We have formerly painted the arid plains of Arabia Petrœa; thofe naked folitudes, where man has never tafted the coolnefs of the fhade, where the fcorched earth, never refrefhed by rain or dew, refts torpid and denies all fubfiftence to every fpecies of being. To this picture of extreme drynefs in the Ancient Continent, let us oppofe the vaft deluged Savannas of the New World. Immenfe rivers, fuch as the Amazons, the Plata, the Oronoco, roll their majeftic billowy

 ftreams

ſtreams, and ſwelling over their banks with unchecked licence, they threaten to uſurp the whole of the land. Sheets of ſtagnant water, widely ſpread, cover their ſlimy ſediment, and theſe vaſt marſhes exhale denſe ſickly vapours, which would poiſon the air, were they not diſperſed by the winds, or precipitated in deſcending torrents. And theſe meadows, which are alternately dry and wet, where the earth and the water ſeem to diſpute their undecided limits, are inhabited only by loathſome animals which multiply in theſe ſewers of nature, where every thing exhibits the image of the monſtrous depoſitions of the primæval ſediment. Enormous ſerpents trace their waving furrows on the miry ſoil; crocodiles, toads, lizards, and a thouſand reptiles of hideous forms crawl and welter in the mud; and millions of inſects, engendered by warmth and moiſture, heave up the ſlime. And this ſordid aſſemblage of creatures which quickens the ground or darkens the ſky, invites numerous flocks of voracious birds, whoſe confuſed notes, mingled with the croakings of the reptiles, while they diſturb the vaſt ſilence of thoſe frightful wilds, inſpire horror, and ſeem to prohibit the approach of man and of every ſentient being.

Amidſt

Amidſt the diſcordant ſounds of the ſcreaming birds and croaking reptiles, there is heard at intervals a powerful note, which drowns the reſt, and rebellows from the diſtant ſhores: it is the cry of the Kamichi, a large black bird, diſtinguiſhed by its voice and its armour. On each wing, it has two ſtrong ſpurs, and on its head a pointed horn * three or four inches long, and two or three lines in diameter at the baſe: this horn, which is inſerted in the top of the forehead, riſes ſtraight, and terminates in a ſharp point bent ſomewhat forward; near the baſe it is ſheathed like the quill of a feather. We ſhall afterwards ſpeak of the ſpurs on the ſhoulders of certain birds, ſuch as the Jacanas, many ſpecies of Plovers, Lapwings, &c. but the Kamichi is by far the beſt armed: for beſides the horn which grows out of the head, it has in each pinion two ſpurs, which project forward when the wing is cloſed. Theſe ſpurs are the apophyſes of the metacarpal bone, and riſe from the anterior part of

* The ſavages of Guiana have named it *Kamichi*; thoſe of Brazil call it *Anhima*, and on the river Amazons *cahuitahu*, in imitation of its powerful cry, which Marcgrave denotes more preciſely by *Vyhu-vyhu*, and which he repreſents as ſomething terrible.

theſe extremities; the upper ſpur is largeſt, of a triang lar form, two inches long, and nine lines broad at the baſe, ſomewhat curved, and terminating in a point; it is alſo inveſted with a ſheath of the ſame ſubſtance as the baſe of the horn. The lower apophyſis of the metacarpal bone is only four lines long, of the ſame breadth at its origin, and ſimilarly ſheathed.

With this furniture of offenſive arms, which would render it formidable in combat, the Kamichi never attacks other birds, but wages war only againſt reptiles. It ſeems even to have a gentle and a feeling diſpoſition; for the male and female keep conſtantly together. Love binds their affections by an indiſſoluble chain: if one happens to die, the ſurvivor can hardly ſupport the loſs of its companion; it wanders perpetually moaning, and conſumes the wretched remainder of its life near the ſcenes of tender recollection and of paſt joys †.

If the affectionate character of this bird forms a contraſt to its mode of life, a like oppoſition obtains in its phyſical ſtructure.

* *Unâ mortuâ, altera a ſepulturâ nunquam diſcedit.* Marcgrave . . . *Raro ſola incedit; verum junctim, mas & fæmina. Teſtantur omnes pariter incolæ, unâ mortuâ alteram inſtar turturum lugere, & vix a ſepulchro diſcedere.* Piſo. *Hiſt. Nat. Ind.* p. 91.

It

It ſubſiſts on prey, and yet it has the bill of a granivorous bird; and, notwithſtanding its ſpurs and its horn, the head reſembles that of a gallinaceous bird; its legs are ſhort, but its wings and tail are very long; the upper mandible projects over the lower, and bends ſomewhat at the point; the head is clothed with ſmall downy feathers, briſtled and criſped, intermixed with black and white; the ſame curled plumage covers the top of the neck; the lower part is covered with broader, thicker feathers, black at the edge, and gray within; all the mantle is browniſh-black, with greeniſh reflections, and ſometimes mixed with white ſpots; the ſhoulders are marked with rufous, and that colour extends on the edge of the wings, which are ſpacious ‡; they reach almoſt to the end of the tail, which is nine inches long; the bill is two inches long, eight lines broad, and ten thick at its baſe; the leg joins to a ſmall naked part of the thigh, and is ſeven inches and a half high, it is covered with a rough and black ſkin, whoſe ſcales are ſtrongly marked on the toes, which are very long; the middle one, including the nail, is five inches, and the nails are half-hooked, ſcooped out below, the hind one being of a peculiar form,

‡ *Alas ampliſſimas.* Marcgrave.

ſlender, almoſt ſtraight, and very long, like that of the Lark. The total length of the bird is three feet: we could not perceive the great difference in point of ſize noticed by Marcgrave between the male and female. Several of the birds which we examined appeared to be nearly of the magnitude of the Turkey cock.

Willughby juſtly remarks, that the Kamichi is the ſole ſpecies of its genus. Its form is compoſed of diſcordant parts, and nature has beſtowed extraordinary qualities on it: the horn which projects from its head would alone diſcriminate it from the whole claſs of birds *. Barrere was miſtaken therefore in ranging it with the Eagles †; ſince it has neither their bill, their head, nor their feet. Piſo ſays juſtly, that the Kamichi is half an Aquatic Bird ‡: he ſubjoins, that it builds its neſt like an oven at the foot of ſome tree, and that it walks with its neck erect and its head lofty, and haunts the foreſts. Yet ſeveral travellers have aſſured us that it is oftener found in the Savannas. [A]

* *Frequens pecora cornuta; raro in aere avem cornua gerentem videris.* Piſon.

† *Aquila Aquatica Cornuta.*

‡ *Rapina eſt & amphibia.*

[A] Specific character of the *Palamedea Cornuta*: "Each pinion has two ſpurs, its front is horned."

N°. 180

THE COMMON HERON.

The COMMON HERON.

First Species.

Ardea Major. Linn. and Gmel.

{ *Ardea.*
{ *Ardea Cristata.* Briss.

Ardea Cinerea. Lath. Gesn. Aldr. Johnst. & Sibb.

Ardea Subcærulea. Schwenckf.

The Common Heron, or *Heronshaw.* Will.

{ *The Crested Heron.*
{ *The Common Heron.* Pennant *.

HAPPINESS is not equally bestowed on all sentient beings. That of man arises from his complacency of mind, and the proper employment of his moral faculties: that of the lower animals results, on the contrary, from their physical qualities, and from the exercise of their corporeal strength. But if nature revolts at the unequal allotment that prevails in human society, she has herself advanced with rapid

* In Greek Ερωδιος or Ερωδας, derived, according to Suidas, from Ελωδης, *belonging to a marsh:* and hence the Latin *Ardea* or *Ardeola:* In Hebrew *Schalach:* In Chaldean *Schalenuna:* In Arabic *Babgach:* In Persian *Aukoh:* In Turkish *Balakzel:* In Hungarian *Cziepie:* In Polish *Czapla* and *Zoraw:* In Italian *Airone, Sgarza:* In Spanish and Portugueze, *Garza:* In German *Reiger:* In Swiss *Reigel:* In Flemish *Reigher:* In Swedish *Haeger:* In Danish *Heyne;* In Norwegian *Heger* or *Kegger.*

rapid ſtrides in the ſame path of injuſtice; and by the imperfection of the organs beſtowed on ſome of her creatures, ſhe has condemned them to ſtruggle perpetually with miſery and want. Neglected children, ſent naked into the world, to live in continual penury; their toilſome days are ſpent in perpetual ſolicitude, ſickened by the reſtleſs cravings of a famiſhed appetite: to ſuffer and to endure, are often their only reſources, and this inward pain traces its ghaſtly impreſſion on their external figure, and obliterates all the graces which attend on felicity. The Heron preſents to us the picture of wretchedneſs, anxiety, and indigence; and as it can procure its prey only by lying in ambuſh, it remains whole hours, whole days in the ſame ſpot, and ſo perfectly ſtill, as to diſcover no ſigns of life. When obſerved by a teleſcope (for it can ſeldom be approached) it appears benumbed, ſeated on a ſtone its body almoſt erect, and reſting on a ſingle foot; its neck folded back along its breaſt and belly, its head and bill ſunk between its ſhoulders, which riſe, and much over-top the breaſt. If it changes its poſture to put itſelf in motion, it aſſumes another, which is ſtill more conſtrained; it advances into the water to the height of its knee, and, holding its head between its legs, it watches the paſſing frog

or

or fiſh. But as it muſt wait the ſpontaneous occurrence of its prey, and has only a moment to make the ſeizure, it is often conſtrained to ſuffer long faſtings, and ſometimes to periſh of abſolute hunger: for when the waters are bound with ice, it has not the inſtinct to retire into milder climates. Some naturaliſts are miſtaken in ranking it among the birds of paſſage * ; for here we ſee Herons at all ſeaſons, and even during the ſevereſt and moſt continued froſts. Then they are compelled to quit the frozen marſhes and rivers, and to repair to the rivulets and perennial ſprings; and at this time they are moſt in motion, and make conſiderable flittings, though only to different parts of the ſame country. They ſeem therefore to multiply, as the ſeaſon grows chiller; and, by dint of patience and ſobriety, they endure equally both hunger and cold. But theſe frigid virtues are commonly attended with a diſguſt for life. When a Heron is caught, it may be kept a fortnight, without ſhewing the leaſt deſire of food, which it even rejects when crammed in its throat; its native gloom, cheriſhed doubtleſs by captivity, ſmothers the ſtrongeſt inſtinct implanted in animated beings, that of ſelf-preſervation: it wears

* *Agricola.*

out

out its existence in complete apathy, without venting a complaint, or betraying the least symptom of tender regret †.

Insensibility, neglect of self-preservation, and some other negative qualities, characterize it better than its positive properties. Pensive and lonely, except in the breeding season, it seems to taste no pleasure, nor even to possess the means of avoiding pain. In the worst weather it remains solitary and exposed, seated on a stake, or on a stone, beside the brink of a rivulet, or on a little eminence in a deluged meadow; while the other birds seek a cover among the foliage: and in the same spots where the Rail shelters itself in the thick herbage, and the Bittern amidst the reeds, the miserable Heron stands unprotected from the violence of the tempest, or the piercing severity of the cold. M. Hebert tells us, that he caught one which was half frozen, and entirely incrusted with ice: he avers also, that he has often found the snow or the mud marked with the prints of their feet, but could never follow the traces more than twelve or fifteen paces; a proof of the narrow compass of their quest, and of

† Experiment made by M. Hebert, to whose excellent observations we owe the principal facts of the natural history of the Heron.

their inaction in the most urgent occasions. Their long legs are only stilts, unfit for running; they hold themselves in an erect posture, and perfectly still the greatest part of the day, and this rest serves instead of sleep; for they fly a little in the night *; then they are heard screaming in the air at all hours, and in all seasons: their cry consists of a single dry sharp note, which might be compared to that of a Goose, if it were not shorter, and somewhat plaintiff †: it is repeated almost incessantly, and prolonged by a shriller and more disagreeable tone when the bird struggles with pain and adversity.

To the hardships necessarily attendant on its toilsome life, the Heron adds ills of its own creation, fear and distrust. Every thing disconcerts and alarms it: it flies from a man at a vast distance; and as it is often attacked by an Eagle or Falcon, it endeavours to escape their assault by rising into the air, and vanishes with its pursuers in the region of the clouds ‡. It was enough

* The ancients had observed this: Eustathius remarks, at the tenth book of the Iliad, that the Heron fishes at night.

† The Greeks expressed the cry of the Heron, from the time of Homer, by the verb κλαιζω. *Iliad x.*

‡ It is asserted, that when driven to the last extremity, it passes its head under its bill and presents its pointed bill to its assailant, which darting with impetuosity, is itself transfixed. Belon. *Nat. des Oiseaux, p.* 190.

that

that nature conſtituted theſe too formidable enemies, and man might have forborne to whet, againſt the unhappy Heron ||, their inſtincts, and inflame their antipathies: yet the chaſe of that bird was once the moſt illuſtrious in falconry; it was the ſport of Princes, who reſerved for themſelves its lean carcaſe as honourable game, qualified by the name of *Royal Meat*, and ſerved up for ſhew at their banquets §.

Hence undoubtedly the pains that have been taken to ſettle Herons in foreſts, or even in towers, commodious eyries being provided for their neſtling. Some profit was drawn from the ſale of the young Herons, which were fatted. Belon ſpeaks with raptures of the Heronries which Francis I. cauſed to be conſtructed at Fontainebleau, and of the aſtoniſhing effect of art which had reduced ſuch ſavage birds under the dominion of man ¶. But the ſucceſs reſulted from their natural diſpoſitions; they love to breed together, and

|| The ancients added many other croſſes to its lot; the Lark broke its eggs, the Woodpecker killed its young; all were its enemies, but the Crow. *See* Ariſtotle, *Lib. ix.* 18 *et* 2.—Pliny, *Lib. x.* 96.

§ *See* Jo. Bruyerinus, *de re cibariâ*, Lib. xv. 56.—Aldrovandus, *tom. iii.* 367. "It is commonly ſaid, that the Heron is royal meat, on which the French nobility ſet great value." Belon. *Nat. des Oiſeaux*, *p.* 190.

¶ *Nat. des Oiſeaux*, *p.* 189.

for

for that purpose multitudes assemble in the same district of the forest * and often on the same tree: and we may suppose that they adopt this measure, to repel the Kite and Vulture by their combined force, or at least, to intimidate them by their numbers. It is on the tallest trees that the Herons build their nests, and often beside those of the Crows † which might have led the ancients to imagine that an amity subsisted between these two species, whose habits and instincts are so incongruous. Their nests are spacious, built of sticks, with abundance of dry grass, of rushes ‡, and feathers; their eggs are of a greenish blue, which is uniformly pale, nearly as large as those of the Storks, but rather longer, and almost equally thick at both ends. The hatch, we are assured, consists of four or five eggs, which ought to make the species more numerous than it appears to be; many must perish therefore in the rigors of winter, and perhaps, as they are of a gloomy disposition and ill fed, they soon lose the power of procreating.

* There is no country where the Herons do not affect certain woods, where they collect together, and make natural *heronries*. They assemble not only on the great oaks, but also on the pines, as Schwenckfeld remarks of some forests in Silesia.

† Aldrovandus, and Belon.

‡ Aristotle, *Lib. ix. 2.*

The

The ancients, impreſſed probably with an idea of the miſerable life of the Heron, imagined that it ſuffered pain even in the act of copulation; and that the male at the critical moment diſcharged blood at the eyes, and ſcreamed with agony §. Pliny drew from Ariſtotle this falſe notion which Theophraſtus countenances: it was refuted however as early as the time of Albertus, who aſſures us that he frequently witneſſed the coition of Herons, and that he perceived nothing but the dalliance of love, and the criſis of pleaſure. The male firſt ſets one foot on the back of the female, to prepare her for the embrace; then carrying both feet forward he ſinks upon her, and holds himſelf in that poſture by ſlightly flapping his wings ||. When ſhe hatches, her mate fiſhes, and ſhares with her his captures; fiſh are often ſeen which have dropt from their neſts *. It does not appear that they

§ "*Ardeolarum . . . pellos in coitu anguntur; mares quidem eum vociferatu ſanguinem etiam ex oculis profundunt; necminus ægre pariunt gravidæ.*" Plin. *Lib. x.* 79. This fable of the torture which the Heron endures in coition has occaſioned another, that of its great chaſtity. Glycas reports that the bird grieves forty days on the proſpect of the ſeaſon of copulation.

|| Johnſton.

* In Low Britany, the Herons are very frequent, where they neſtle in the foreſts in the tall trees; and as they feed their

they feed on ſerpents or other reptiles. We know not why in England it is prohibited to kill this bird †.

We have ſeen that the adult Heron rejects food and dies of hunger: but, if caught young it may be tamed, reared and fattened. We have ſeen ſome carried from the neſt to the court-yard, where they lived on fiſh guts, and raw fleſh, and aſſociated with the fowls. They are ſuſceptible, not indeed of education, but of certain communicated movements; they have been taught to wreath their neck in different faſhions, and to entwine it about their maſter's arm. But if they were not ſtimulated, they ſoon relapſed into their natural melancholy, and remained ſtill and fixed ‡. The young Herons are at firſt covered, for a conſider-

their young with fiſhes, many of theſe fall to the ground: ſeveral perſons have therefore taken occaſion to ſay, that they had been in a country where the fiſh which dropt from the trees, fatten the ſwine." Belon, *Nat. des Ois. p.* 189.

† *Ardeam in Angliâ occidere capitale eſſe ferunt.* Muſ. Worm. Johnſon ſays the ſame. [This matter ſeems to be exaggerated. It was formerly reckoned in this country a bird of game, and to preſerve the ſpecies, the breaking of the eggs was ſubject to a penalty of twenty ſhillings. T.]

‡ "I kept one in my court; it did not ſeek to eſcape, it never fled when approached, but remained motionleſs where it was ſet: the firſt days, it preſented its bill and even ſtruck with the point, but without hurting. I never ſaw an animal more patient, more motionleſs, and more ſilent." *M. Hebert.*

 able

able time, with a thick hairy down, chiefly on the head and the neck.

The Heron catches numbers of frogs, and ſwallows them entire; this fact is aſcertained from its excrements, which contain their bones cruſhed and enveloped in a viſcous mucilage of a green colour, formed moſt probably of the frogs ſkins reduced to glue. Its excrements, like thoſe of the Aquatic Birds in general, ſhew an acrid quality on herbs. In times of ſcarcity, it ſwallows ſome ſmall plants, ſuch as the water lentil *; but its ordinary food is fiſh. It catches very ſmall ones, and muſt neceſſarily have a nice and prompt aim to ſtrike prey that glances with ſuch rapidity. With regard to large fiſhes, however, Willughby ſays, with great probability, that it pecks and wounds much more than it draws out of the water. In winter when the froſt generally prevails, and it is obliged to reſort to the tepid ſprings, it wades feeling with its foot in the mud, and thus diſcovers its prey, whether it be a frog or a fiſh.

By means of its tall legs, the Heron can enter into water more than a foot deep without wetting itſelf: its toes are extremely long, the middle one being equal to the

* Salerne.

Tarſus,

Tarſus, and the nail which terminates it is indented within, like a comb †, and ſerves to ſupport it, and to cling to the ſlender roots entangled in the mud. Its bill is jagged with points turned backwards, which ſecure the fiſh from ſlipping out of its hold. Its neck often bends double, and this motion would ſeem to be performed by a hinge; for it can be bent ſo, ſeveral days after the bird is dead. Willughby has falſely aſſerted, that the fifth *vertebra* is reverſed; ſince on examining the ſkeleton of a Heron, we counted eighteen *vertebræ*, and of theſe we remarked, that only the five firſt were ſomewhat compreſſed at the ſides, and jointed one to another by a projection of each preceding upon the next following, without apophyſes, which did not begin to appear till on the ſixth *vertebra*. By this ſingular ſtructure, the part of the neck adjacent to the breaſt is ſtiffened, and that contiguous to the head plays in a ſemicircle on the other, and applies to it in ſuch a manner, that the neck, the head, and the bill, are folded in three pieces, one upon another. The bird ſuddenly, as if by the action of a ſpring,

† This comb-like indenting is carved on the dilated and protuberant edge of the inſide of the nail, without extending ſo far as the point which is ſharp and ſmooth

 extends

extends this doubled portion, and darts its bill like a javelin By ſtretching out its neck to the full length, it can reach at leaſt three feet all round: and, when at reſt, the neck almoſt diſappears, and is hid between the ſhoulders, to which the head ſeems attached ‡. Its cloſed wings project not beyond the tail, which is very ſhort.

To fly, it extends its legs ſtiff backwards, bends its neck upon its back, and folds it into three parts, including the head and bill; ſo that from below the head is not viſible, and the bill appears to protrude from the breaſt. It diſplays larger wings than any bird of prey, and they are concave, and ſtrike the air with an equal and regular motion: and by this uniform flight, the Heron is enabled to ſoar ſo high as to be loſt in the clouds §. It flies ofteneſt before rain ||; and from its actions and poſtures, the ancients drew conjectures with regard to the ſtate of the air and the changes of the weather. If it remained ſtill and forlorn on the beach, it foreboded wintery cold *; if more than ordinarily reſtleſs and clamorous, it promiſed rain; and if its head

‡ Willughby.

§ ———————— *notaſque paludes*
Deſerit, atque altam ſupra volat ardea nubem.
Virg. *Georg. i.* 363.

|| Aldrovandus.

* *Ardea in mediis arenis triſtis, hiemem.* Plin. *Lib. lxiii.* 87.

reſted

reſted on its breaſt, the direction of the bill ſignified the quarter from which the wind was to blow †. Aratus and Virgil, Theophraſtus and Pliny, eſtabliſh theſe indications, which have been neglected as uſeleſs, ſince means of art more certain have been dicovered.

Few birds ſoar ſo high as the Herons, or traverſe ſuch extenſive tracts in the ſame climate: and often, ſays M. Lottinger, ſome are caught which bear the marks of the places where they haunted. Indeed it muſt require ſmall force to tranſport ſo ſlender and meager a body, which is ſhrunk, alſo flat at the ſides, and much more covered with feathers than with fleſh. Willughby imputes the leanneſs of the Heron, which is exceſſive, to the perpetual fear and anxiety which torture it, as well as to the want and inactivity of its condition ‡.

 The

† Aldrovandus.

‡ " I fired at a Heron, when the weather was exceſſive cold; the bird was only ſlightly wounded, and bore the ſhot to a good diſtance. A large dog which I had with me, and which was vigorous and bold, heſitated to run in upon this Heron, till he perceived me near him: the bird ſcreamed frightfully, turned on its back, and preſented its feet to ſhield it when one approached nigh; yet when I took it up, though it was full of life and of ſtrength, it did not hurt me, nor even attempt any injury. I ſkinned it for a preparation; it was exceſſively lean. I had ſurprized it early in

The whole genus ef the Herons have, like the quadrupeds, only one *cæcum*; whereas, in all other birds in which that gut is found, it is double §: the *œsophagus* is very broad and capable of a great dilatation: the *trachea arteria* is sixteen inches long, contains about fourteen rings, and is nearly cylindrical as far as its bifurcation, where it swells considerably and sends off two branches, which internally consist of only one membrane: the eye is placed in a naked greenish skin, which extends as far as the corners of the bill; the tongue is pretty long, soft, and pointed: the bill is cleft up to the eyes, and discovers a long wide aperture; it is strong, thick near the head, six inches long, and terminating in a sharp point: the lower mandible is sharp at the edges, near three inches long, and hollowed by a double groove, in which the nostrils are placed; it is of a yellowish colour and brown at the point: the lower mandible is yellower, and the two branches that compose it, do not join till within two inches

in the morning on the brink of a very deep river, where it certainly could not make frequent captures; and several days ago I met with Herons at the same place, as I was seeking for Wild Ducks." *Note extracted from the excellent Memoir of* M. Hebért *on Herons.*

§ Aristotle was little acquainted with the Heron, when he said that it was active and crafty in procuring its food.

inches of the tip; the interſtice between them is furniſhed with a membrane covered with white feathers: the throat too is white, and beautiful black ſtreaks mark the long feathers which hang on the foreſide of the neck. All the upper ſide of the body is of a fine pearl gray; but, in the female, which is ſmaller than the male, the colours are paler, not ſo deep or gloſſy; the black croſs bar on the breaſt is alſo wanting, and the tuft on the head: in the male, there are two or three long ſprigs of thin, ſlender, flexible feathers of the fineſt black; theſe feathers are highly prized in the Eaſt *. The tail of the Heron contains twelve quills, in the ſlighteſt degree tapered: the naked part of the thigh is three inches; the tarſus ſix; the great toe above five, and is joined to the inner toe by a portion of a membrane; the hind toe is alſo very long, articulated with the outer, and inſerted into the ſide of the heel, a ſingular property which obtains in all birds of this family: the toes, the legs, and the thighs, of the Common Heron, are of a greeniſh yellow: it meaſures five feet acroſs the wings, and near four from the

* Klein.—There are three famous plumes compoſed of theſe rare Herons' feathers; that of the Emperor, that of the Grand Turk, and that of the Mogul; but if theſe feathers be really white, as it is alledged, they muſt belong to the Night Heron.

tip of the bill to the nails, and a little more than three to the end of the tail; its neck is ſixteen or ſeventeen inches, and when it walks, it carries more than three feet of height: it is therefore almoſt as large as the Stork, but of a much thinner body, ſince, notwithſtanding its bulk, it weighs ſcarce more than four pounds †.

Ariſtotle and Pliny ſeem to have known only three kinds of Herons: the Common or Large Gray Heron, which we have juſt deſcribed, and which they termed *Pellos* or the cinereous; the White Heron, or *Leucos*; and the Stellated Heron, or *Aſterias* ‡. Yet Oppian remarks, that the ſpecies of the Heron are more numerous and varied. In fact, each climate has ſome appropriated to itſelf, as we ſhall perceive from the enumeration; and the Common Gray Heron ſeems to have penetrated into almoſt all countries, and ſettled with the indigenous kinds. No ſpecies indeed is ſo ſolitary or ſo ſcanty in each country; but, at the ſame time, none is more widely diſperſed, or ſcattered more remotely in oppoſite climates. Its phlegmatic temper and its laborious life,

† A male Heron, taken the 10th of January, weighed three pounds ten ounces; a female, three pounds five ounces. *Obſervation made by M. Gueneau de Montbeillard.*

‡ Ariſt. *Lib. ix.* 2. Plin. *Lib. x.* 79.

have

have reconciled to every viciſſitude. Dutertre aſſures us, that among the multitude of Herons peculiar to the Antilles, the European Gray Heron often occurs. It has been found even in Otaheite, has an appellation in the language of that iſland *, and like the King fiſher, it is venerated by the natives. In Japan, we diſtinguiſh, ſays Koempfer, amidſt many ſpecies of *Saggis* or Herons, the *Goi-ſaggi* or Gray Heron. It is met with in Egypt †, in Perſia ‡, in Siberia, and in the territories of the Jakutes §. The ſame ſpecies appears in the iſland of St. Jago; at Cape Verd ||; in the Bay of Saldana **; in Guinea ††; in the iſle of May ‡‡; in Congo §§; in Guzarat ||||;

* *Otoo.*

† Voyage de Granger; *Paris*, 1745, *p.* 237. — Voyage de P. Vanſleb; *Paris*, 1677, *p.* 103.

‡ Voyage de Chardin; Amſterdam, 1711, *tom. ii. p.* 30.

§ Gmelin; *Hiſt. Gen. des Voyages, tom. xviii. p.* 300.

|| Hiſt. Gen. des Voyages *tom. ii. p.* 376.

** Idem, tom. i. p. 449.

†† "Here are found (on the Gold Coaſt) two kinds of Herons, blue and white." *Voyage en Guinee* par Guillaume Boſman, *Utrecht*, 1705.

‡‡ *See* Roberts' account in *Hiſt. Gen. des Voyages*, tom. ii. p. 37.

§§ Beſides the birds which are peculiar to the kingdom of Congo and of Angola, Europe has few that occur not in one or other of theſe two countries: Loppez obſerves that the pools are filled with Herons and Gray Bitterns, which are ſtyled *Royal Birds. Hiſt. Gen. des Voyages, tom. v. p.* 75.

|||| Mandeſlo in Olearius tom. ii. p. 145.

in

in Malabar*; in Tonquin †; in Java ‡, and in Timor §. The Heron called *Dangcanghac* in the iſland of Luçon, and to which the ſettlers in the Philippines apply the Spaniſh name for a Heron, *Garza*, appears to be the ſame ‖. Dampier expreſly ſays, that the Heron of the Bay of Campeachy is exactly like that of England; which, joined to the teſtimony of Dutertre, and Dupratz who ſaw the European Heron in Louiſiana, leaves no room to doubt that it is common to both continents; though Cateſby aſſerts, that all thoſe which he found in the New World were of a different ſpecies.

Diſperſed and ſolitary in the inhabited countries, the Herons are collected and numerous in ſome deſert iſlands, as in thoſe in the Gulph of Arguim at Cape Blanc, which, for that reaſon, the Portugueſe have termed *Iſola das Garzas*: the eggs were found in ſuch quantities as to load two boats ¶. Aldrovandus ſpeaks of two iſlands on the African coaſt which received the

* Recueil des Voyages qui on ſervi à l'établiſſement de la Compagnie des Indes; *Amſterdam*, 1702, *tom. vi. p.* 479.

† Dampier.

‡ Nouveau Voyage autour du monde, par le Gentil, tom. iii. p. 74.

§ Dampier.

‖ *See* Camel. *Philoſ. Tranſact.* No. 288.

¶ Cadamoſto, *Hiſt. Gen. des Voyages.* tom. ii. p. 291.

ſame

ſame name from the Spaniards. The iſlet of the Niger, where Adanſon landed, merited a like appellation *. In Europe, the Gray Heron has penetrated to Sweden †, Denmark, and Norway ‡. It is ſeen in Poland §, in England ||, and in moſt of the provinces in France. It is moſt abundant in countries interſperſed with ſtreams and marſhes, ſuch as Switzerland ** and Holland ††.

We ſhall divide the numerous genus of the Herons into four families: that *of the Heron properly ſo called*; that of the *Bittern*; that of the *Bihoreau* or Night Heron; and that of the *Crab Heron*. Their common characters are the length of neck; the ſtraightneſs of the bill, which is pointed in the upper mandible and indented at the edges near the tip: the length of the wings, which,

* "We arrived on the 8th at Lammai (a little iſland in the Niger;) the trees were covered with ſuch a prodigious multitude of Cormorants and of Herons of all kinds, that the Laptots who entered into a ſtream which we then croſſed, filled in leſs than half an hour a canoe, with the young ones, which were caught by the hand, or knocked down with ſticks, and with old ones, of which ſeveral dozens fell at each ſhot of the fowling piece. Theſe birds taſted of fiſh-oil, which does not ſuit every body's palate."

† *Fauna Sueuca*, No. 133.

‡ Brunnich, *Ornithol. Boreal.* No. 156.

§ Rzaczynſki.

|| Natural Hiſtory of Cornwall.

** Geſner.

†† Voyage hiſtorique de l' Europe: *Paris*, 1693, *tom.* v. *p.* 73.

when

when closed, cover the tail; the height of the tarsus, and of the naked part of the thigh; the great length of the toes, of which the middle one has its nail indented, and the singular position of the hind one, which is articulated at the side of the heel near the inner toe; lastly, the naked greenish skin which spreads from the bill to the eyes in them all. Their habits also are nearly the same; for they all haunt the marshes and margins of water; they are patient, slow in their motions, and have a melancholy deportment.

The peculiar features of the family of Herons, in which we include the Egrets, are, the neck excessively long, very slender, clothed below with pendant loose feathers; the body strait, shrunk, and, in most species, set on tall stilts.

The Bitterns have a thicker body, and not so tall legs as the Heron; their neck is shorter, and so well clothed with feathers, as to appear very thick in comparison with that of the Heron.

The *Bihoreaus*, or Night Herons, are not so large as the Bitterns; their neck is shorter, and two or three long feathers inserted in the nape of the neck distinguish them from the three other families; their upper mandible is slightly arched.

The

The Crab Herons, which may be termed the ſmall Herons, form a ſubordinate family, and are nothing but a repetition of the Herons on a ſmaller ſcale. None of them equal the Egret, which has only one quarter of the bulk of the Common Heron. The *Blongios*, which exceeds not a Rail, terminates the extenſive range of ſpecies, of a genus more than any other varied both in ſize and form. [A]

[A] Specific character of the Creſted Heron of Pennant, *Ardea Major*. " On the back of the head is a black pendulous creſt, its body is cinereous; a line on the underſide of its neck, and a bar on its breaſt, black." To this ſpecies Linnæus refers as a variety the Common Heron of Pennant, *Ardea Cinerea*, thus characterized ; " the back of its head is black and ſmooth, its back bluiſh, below whitiſh, with oblong black ſpots on its breaſt." Mr. Latham, with the illuſtrious Buffon, very properly ſtates the former as the male, and the latter as the female.

The Heron is ſaid to live to a great age : Keyſler mentions one above ſixty years old.

The GREAT WHITE HERON.

Le Heron Blanc. Buff.

Second Species.

Ardea Alba. Linn. Gmel. Geſner. Aldrov. and Johnſt.

Ardea Candida. Briſſ. and Schwencf.

Ardea Alba Major. Ray. Will. Brown.

The Great White Gaulding. Brown and Sloane *.

As the ſpecies of Herons are numerous, we ſhall ſeparate thoſe of the Ancient Continent, which amount to ſeven, from thoſe of the New World, of which we already know ten. The firſt of the ſpecies that inhabits our continent, is the Common Heron juſt deſcribed; the ſecond is the White Heron, which received that appellation from Ariſtotle. It is as large as the Gray Heron, and its legs are even taller; but it wants the tufts, and ſome naturaliſts have inaccurately confounded it with the Egret †. All its plumage is white, its bill is yellow, and its legs are black. Turner ſeems to allege, that the White Heron has been ſeen to copulate with the gray:

* In Greek Ερωδιος λευκος, or Λευκερωδιος: in Latin *Ardea Alba*, or *Albardeola*: in Italian *Garza Bianca*: and in German *Weiſſer Reiger*.

† Salerne.

but

but Belon, with more probability, ſays only that the two ſpecies conſort, and live in ſuch friendſhip as often to rear their young in the ſame eyry. It appears therefore that Ariſtotle was miſinformed, when he aſſerted that the White Heron conſtructs its neſt with more art than the Gray Heron *.

Briſſon gives a deſcription of the White Heron †, to which we muſt add that the naked ſkin about the eyes is not entirely green, but mixed with yellow on the edges; that the iris ‡ is lemon-colour; and that the naked part of the thighs is greeniſh.

The White Herons are frequent on the coaſts of Britany §; and yet the Species is very rare in England ||, though common in the North, as far as Scania **. They are not ſo numerous as the Gray Herons††. but equally diſperſed: for they are found

* Hiſt. Animal. *Lib. ix.* 24.

† "Its whole body white; a naked green ſpace between its bill and its eyes; its bill ſaffion yellowiſh; its legs black." *Briſſ.*

‡ Extract of a letter from Dr. Hermann to M. de Montbeillard, dated Straſburg, 22d September 1774.

§ Belon.

|| Britiſh Zoology.

** *Fauna Suecica.*

†† Schwenckfeld.

in

in New Zealand *, in Japan †, in the Philippines ‡, at Madagascar §, in Brazil ||, where it is called *Guiratinga*, and at Mexico **, under the name of *Aztatl*.

* "One of the party shot a White Heron, which agreed exactly with Mr. Pennant's description, in his British Zoology, of the White Herons that either now are, or were formerly in England." In the language of the Society-isles, the White Heron is called *tra-pappa*. Cook's Second Voyage, *vol.* 1. *p.* 87.

† It is there called *Suro-saggi*, according to Kœmpfer.

‡ Camel, *Philosophical Transactions*, No. 285.

§ In the language of that island it is termed *Vahon-vahon-fouchi*, according to Flaccourt.

|| *Hist. Nat. Brasil*, p. 210. De Laet describes the *Guiratinga* in terms that exactly define the White Heron. "The *Guiratinga* is one of the birds which live in the sea; it is as large as a Crane, its feathers bright white; its bill long and sharp, of a saffron colour; its legs long, and of a yellowish red; its neck cloathed with feathers so fine and elegant, that they may be compared to the plumes of the Ostrich." Nov. Orb. *p* 575.

** The Aztatl, or White Heron, has either the same size and shape with our Heron, or very nearly so; the feathers on the whole of the body are snowy, very soft, and wonderfully crisped and clustered; the bill long and palish, near its origin greenish; its legs long and black."

[A] Specific character of the *Ardea Alba*: its head is smooth, its body white, its bill fulvous, its legs black."

The BLACK HERON.

Third Species.

Ardea Atra. Gmel.
Ardea Nigra. Briſſ. Klein and Schwenck.

SCHWENCKFELD would have been the only naturaliſt that ever noticed this Heron, had not the authors of the Italian Ornithology mentioned a Sea Heron, which they ſay is black: and as Schwenckfeld ſaw his in Sileſia, which is far inland, theſe two birds may be different. It was as large as the Common Heron; all its plumage blackiſh, with blue reflections on the wings. It would ſeem, that the ſpecies is rare in Sileſia; yet we may preſume that it is more frequent in other parts, and that it viſits the ſeas; for it probably occurs in Madagaſcar, where it has a proper name *. But we ought not, as Klein has done, to refer to the ſame ſpecies the *Ardea cæruleo-nigra* of Sloane, which is much ſmaller, being the Crab Heron of Labat. [A.]

* *Vahon-vahon-maintchi*, according to Flaccourt.

[A] Specific character of the *Ardea Atra*: "It is all black, its face naked, its head ſmooth."

The PURPLE HERON.

Fourth Species.

{ *Ardea Purpurata.*
{ *Ardea Purpurea.* Linn. and Gmel.

{ *Ardea Purpurascens.*
{ *Ardea Cristata Purpurascens.* Briss.

{ *The Purple Heron.*
{ *The Crested Purple Heron.* Lath.

THE *Purple Heron of the Danube*, described by Marsigli, and the *Crested Purple Heron* of our *Planches Enluminées* seem to be the same. The crest is the attribute of the male and the small differences in their colours may result from age or sex: the bulk too is the same, though Brisson represents his *Purplish Crested Heron*, as much smaller than the Purple Heron of Marsigli. Their dimensions are nearly equal to each other, and to those of the Common Heron: the neck, the stomach, and part of the back, are of a fine purple rufous; long slender feathers of the same colour rise from the sides of the back, and extend to the end of the wings, falling back on the tail.

The VIOLET HERON.

Fifth Species.

Ardea Leucocephala. Gmel.

THIS Heron was ſent to us from the Coaſt of Coromandel: the whole of the body is a very deep bluiſh, tinged with violet; the upper ſide of the head is of the ſame colour, and alſo the lower part of the neck, the reſt of which is white: it is ſmaller than the White Heron, and not more than thirty inches long.

The WHITE GARZETTE.

Sixth Species.

Ardea Æquinoctalis. Var. 1. Gmel.
Ardea Candida Minor. Briſſ.

ALDROVANDUS beſtows on this White Heron, which is ſmaller than the firſt, the names *Garzetta* and *Garza-bianca*; diſtinguiſhing it at the ſame time from the Egret, which he had previouſly characterized. Yet Briſſon has confounded them; he refers Aldrovandus's *Garza-bianca* to the

the Egret, and substitutes in its place, under the appellation of *Little White Heron*, a small species whose plumage is white, tinged with yellowish on the head and breast *, and which appears to be only a variety of the Garzette, or rather the Garzette itself, but younger and with a trace of its early garb, as Aldrovandus shews by the epithets which he applies †. The adult bird is entirely white, except the bill and the legs which are black; it is much smaller than the Great White Heron, not being two feet long Oppian seems to have known this species ‡. Klein and Linnæus make no mention of it, and probably it is not found in the North. Yet the White Heron which Rzaczynski says is seen in Prussia, and whose bill and legs are yellowish, appears to be only a variety of this species; for in the Great White Heron, the bill and legs are constantly black: and this is the more probable, as in France even this small species of Garzette is subject to other varieties. M. Hebert assures us, that in Brie, in the month of April, he killed one of these Small White Herons, which was not larger

* *Ardea Minor alia, vertice croceo.* Aldrov.

† Its body smaller, not so compact, its bill entirely yellow, &c.

‡ "There are some Small and White Herons." *Exeutic.*

than

No. 181

THE EGRET.

than a Field-pigeon; its legs green with a ſmooth ſkin, while in other Herons the legs have commonly a rough mealy ſurface §.

§ "I again ſaw, in 1757, three of the ſame Herons on the margin of the lake of Nantua during an exceſſive cold; they appeared there about eight days, till the lake was frozen over." *Note communicated by M. Hebert.*

The LITTLE EGRET.

L'Aigrette. Buff.

Seventh Species.

Ardea-Garzette. Linn. and Gmel.
Egretta. Briſſ.
Garzetta Italorum. Johnſt. and Charleton.
The Criel Heron. Harris. Coll. Voyag.

BELON is the firſt who gave the name of *Aigrette* (tuft) to this ſmall ſpecies of White Heron, and probably becauſe of the long ſilky feathers on its back; theſe being employed to decorate the ladies head dreſs, the warrior's helmet, and the ſultan's turban. They were in great requeſt formerly in France, when our doughty champions wore plumes. At preſent, they ſerve for a gentler uſe; they deck the heads of our beauties, and raiſe their ſtature: the flexibility,

flexibility, the ſoftneſs, and the lightneſs of theſe feathers, beſtow grace on their motions.

Theſe feathers conſiſt of a ſingle delicate ſhaft, from which at ſmall intervals, riſe pairs of very fine threads, as ſoft as ſilk: on each ſhoulder, there is a tuft of theſe fine feathers, which extend on the back and beyond the tail; they are of a ſnowy white, as well as the other feathers, which are coarſer and harder. Yet the bird, before it firſt moult, and perhaps even later, has a mixture of gray or brown in its plumage. One of thoſe killed by M. Hebert in Burgundy *, had every appearance of being young, and particularly its plumage was ſtained with dark colours.

The Egret is one of the ſmalleſt of the Herons, being commonly not two feet long. When full grown, its bill and legs are black. It prefers the haunts of the ſea-ſhore; yet it perches and neſtles on trees, like the other Herons.

It appears that the European Egret occurs in America †, with another larger ſpecies

* At Magny, on the banks of the Tille, 9th May 1778.

† Dutertre, *Hiſtoire des Antilles, tom. ii. p.* 777, " Among the birds that haunt rivers and pools . . . there are Egrets of an admirable whiteneſs, of the bulk of a Pigeon . . . They are particularly ſought for on account of the precious bunch

ſpecies, which we ſhall deſcribe in the following article. It is diſperſed alſo to the moſt remote and lonely iſlands, ſuch as that of Bourbon ‡, and the Malouines §. It is found in Aſia in the plains of the Araxes ||, on the ſhores of the Caſpian Sea ¶, at Siam **, at Senegal and Madagaſ-

 car,

bunch of feathers, fine and delicate like ſilk, with which they are decorated, and which gives them a peculiar grace." *Hiſt Nat & Moral des Antilles; Rotterdam*, 1658, *p.* 149. Father Charlevoix ſays that there are *Fiſhers* or *Egrets* in St. Domingo, which are real Herons, little different from our own. *Hiſt. de St. Domingue, Paris* 1730, *tom. i.*

‡ Voyage de François Leguat; *Amſterdam*, 1708, *tom.* 1. *p.* 55.

§ "The Egrets are pretty common (at the Malouine or Falkland Iſlands); we took them for Herons, and we knew not at firſt the value of their plumes. Theſe birds begin to fiſh about the cloſe of the day: they bark from time to time, inſomuch that one might believe they were the Wolf-foxes already mentioned." *Voyage autur du Monde*, par M. de Bougainville, *tom.* 1. *p.* 125.

|| Voyage de Tournefort, *tom. ii. p.* 353.

¶ The Heron and the Egret are common round the Caſpian Sea and that of Azoff: the Ruſſians and Tartars know theſe birds, and eſteem them for their precious plumes; the former call them *Tſchapla-belaya*, the latter, *Ak koutan. Diſcourſe on the Commerce of Ruſſia*, by M. Guldenſtaed.

** "Nothing is more charming to ſee than the great number of Egrets, with which the trees are covered (at Siam): at a diſtance, they would ſeem to be the flowers; the mixture of the white of the Egrets with the green of the leaves, has the fineſt effect in the world. The Egret is a bird of the figure of a Heron, but much ſmaller; its ſhape is delicate; its plumage beautiful, and whiter than ſnow; it

car *, where it is called *Langhouron* †. But the Black, Gray, and Purple Egrets, which the navigators Flaccourt and Cauche place in that ſame iſland, may, with the greateſt probability, be referred to the foregoing ſpecies of Herons.

it has tufts on its head, on its back, and under its belly, which conſtitute its principal charms, and render it extraordinary." *Dernier Voyage de Siam*, par le P. Tachard, *Paris*, 1686, p. 201.

* Along the river (Gambia) the Dwarf Heron is found, which the French term *Aigrette*: It reſembles the Common Herons, except that its bill and legs are entirely black, and its plumage pure white; on the wings and the back is a ſort of delicate feathers, twelve or fifteen inches long, which in French are called *Aigrettes*. They are much eſteemed by the Turks and Perſians, who employ them to decorate their turbans." *Hiſt. Gen. des Voyages, tom. iii. p.* 305.

† Flaccourt.

HERONS of the New Continent.

The GREAT EGRET.

Le Grande Aigrette. Buff.

First Species.

Ardea-Egretta. Gmel.

ALL the preceding ſpecies of Herons are natives of the Old Continent; all thoſe which follow belong to the new. They are exceedingly numerous in regions where the water ſpreads unreſtrained over vaſt tracts, and where all the low grounds are deluged. The Great Egret is undoubtedly the moſt beautiful of theſe ſpecies, and never occurs in Europe. It reſembles our Egret in the beautiful white of its plumage, without mixture of any other colour: it is twice as large, and conſequently its magnificent attire of ſilky feathers is the richer and fuller. Like the European Egret, it has its bill and legs black. At Cayenne it breeds in the ſmall iſlands that riſe out of the overflowed Savannas: it does not fre-

quent

quent the margin of ſalt marſhes or of the ſea; but conſtantly haunts the ſtagnant waters and the rivers, and lodges among the ruſhes. It is pretty common in Guiana, though it does not form flocks like the Little Egrets. It is alſo ſhyer, more difficult of approach, and ſeldom perches. It is ſeen in St. Domingo, where, in the dry ſeaſon, it lives beſide the marſhes and pools. Nor is it confined to the hotteſt parts of America, ſince we have received ſome ſpecimens from Louiſiana.

[A] Specific character of the *Ardea Egretta*: "It is ſomewhat creſted white, its legs black, the feathers on its back and breaſt looſe and narrow, its pendants very long."

The RUFOUS EGRET.

Second Species.

Ardea Rufeſcens. Gmel.
The Reddiſh Egret. Penn. and Lath.

THIS Egret has a blackiſh gray body; tufts on the back and ruſty unwebbed feathers on the neck. It is found in Louiſiana, and is not quite two feet long.

The DEMI-EGRET.

Third Species.

Ardea Leucogaſter. Gmel.

WE have in the *Planches Enluminées* termed this bird the *Bluiſh White bellied Heron of Cayenne*: it ſeems to be intermediate between the Egrets and the Herons: for, inſtead of the large bunch of feathers of the Egrets, it has only a tuft of rufous unwebbed ſhoots, which is wanting however in the other Herons. It is not two feet long; the upperſide of the body, the neck and the head are deep bluiſh, and the under ſide of the body is white. [A]

[A] Specific character of the *Ardea Leucogaſter*: "It is dark blue, below white, a double-feathered creſt on the back of its head; its bill, its naked face, and its legs yellow."

The SOCO.

Fourth Species.

Ardea-Cocoi. Linn. and Gmel.
Cocoi Brasiliensibus. Marcg. Johnst. Will. Ray.
Ardea Cayanensis Cristata. Briss.
The Cocoi Heron. Lath.

SOCO, according to Piso, is the generic name of the Brazil Herons; but we shall appropriate it to the large beautiful species, which Marcgrave makes his second species, and which occurs in Guiana and in the Antilles, as well as in Brazil. It is equal in bulk to our Gray Heron; it has a crest, which consists of fine pendant feathers, some of them six inches long, and of an handsome ash-colour; according to Dutertre, the old males alone wear this bunch of feathers; those which hang under the neck, are white and equally delicate, soft, and flexible; those of the shoulders and of the mantle are of a slaty gray. Piso says that this bird is commonly very lean, but remarks however that it grows plump in the rainy season. Dutertre calls it the *Crabeater*, (*Crabier*) as usual in the island, and says that it is not so frequent as the other Herons, but that its flesh is as good, or rather is not worse.

The BLACK CAPPED WHITE HERON.

Fifth Species.

Ardea Pileata. Lath. Ind.
The Black-Crested White Heron. Lath. Syn.

THIS Heron is found in Cayenne. Its whole plumage is white, except a black cap on the crown of its head, which bears a tuft of five or six white feathers; it is scarce two feet long; it inhabits up the rivers in Guiana, but is rare. We shall class with it the White Heron of Brazil; the difference of size being accidental, and the black spot and tuft may, as usual in the Herons, be the attribute of the male.

The BROWN HERON.

Sixth Species.

Ardea Fusca. Lath.

IT is larger than the preceding, and is also a native of Guiana. All the upper side of the body is blackish brown, the tint being

being deeper on the head, and ſhaded with bluiſh on the wings; the forepart of the neck is white, charged with browniſh daſhes; the under ſide of the body is pure white.

The AGAMI HERON.

Seventh Species.

Ardea Agami. Gmel.

WE know not why this Heron, which we received from Cayenne, had the name of *Agami*; unleſs becauſe of the long feathers in the tail of both birds. Theſe feathers are of a deep blue: the underſide of the body is rufous; the neck is of the ſame colour before, but bluiſh below, and dark blue above; the head is black, the occiput bluiſh, from which hang long black filaments.

The HOCTI.

Eighth Species.

Ardea-Hoactli. Gmel.
Hoactii, sive Tolactli. Fernand. and Ray.
Ardea Mexicana Cristata. Briss.
The Dry Heron. Lath.

NIEREMBERG interprets the Mexican name of this bird, *Hoactli* or *Toloactli*, by *Avis Sicca*, dry or lean bird; an appellation which well suits a Heron. The present is only one half the bulk of the Common Heron. Its head is covered with black feathers which extend to the nape in a bunch; the upper side of the wings and the tail are of a gray colour; on the back, there are some black feathers, glossed with green; all the rest of the plumage is white. The female has a different name, *Houcton*; it is distinguished by some colours of its plumage; it is brown on the body, mixed with several white feathers, and white on the neck, interspersed with brown feathers.

This bird is found in the lake of Mexico; it breeds among the rushes, and has a strong hoarse voice resembling that of the Bittern. The Spaniards term it very improperly a Kingfisher, *Martinete Pescador*.

The HOUHOU.

Ninth Species.

Ardea-Houhou. Gmel.
Ardea Mexicana Cinerea. Briss.

WE have formed this name by contraction from the Mexican word *Xoxouquihoactli*, which is pronounced *Hohouquihoactli*, *Houhou* represents its cry: Fernandez, who mentions it, subjoins that it is a small species; yet its length is *two cubits*. The belly and neck are cinereous; the front is white and black; the crown of the head and the tuft on the occiput are of a purple colour, and the wings are variegated with gray and bluish. This bird is rare; it is seen from time to time on the lake of Mexico, where it probably arrives from the more northern countries.

The GREAT AMERICAN HERON.

Tenth Species.

Ardea-Herodias. Linn. and Gmel.
Ardea Virginiana Criſtata. Briſſ. and Klein.
The Largeſt Creſted Heron. Cateſby.
The Great Heron. Penn. and Lath.

OF the Marſh birds, the largeſt and the moſt numerous ſpecies occur in the New World. The Heron found by Cateſby in Virginia is the largeſt in the genus: it is near four feet and a half high when erect, and meaſures almoſt five feet from the bill to the nails; its bill is ſeven or eight inches long; all its plumage is brown, except the quills of the wing, which are black; it has a creſt of brown unwebbed feathers. It feeds not only on fiſh and frogs, but alſo on large and ſmall lizards. [A]

[A] Specific character of the *Ardea Herodias*: "The back of its head is creſted, its body brown, its thighs rufous, its breaſt marked with oblong black ſpots.

The HUDSON'S BAY HERON.

Eleventh Species.

Ardea Hudsonias. Linn. and Gmel.
Ardea Freti Hudsonis. Briss.
The Ash-coloured Heron. Edw.
The Red-shouldered Heron. Penn. and Lath.

THIS Heron too is very large; it is near four feet from the bill to the nails, a fine crest of black brown, disposed behind, shades its head; its plumage is light brown on the neck, deeper on the back, and still browner on the wings; the shoulders and thighs are of a reddish brown; the stomach is white, and also the great feathers which hang from the foreside of the neck, which are dashed with brown streaks.

These are all the species of Herons known to us; for we do not admit into their number the eighth species described by Brisson from Aldrovandus; because the bird was young, and had still its first garb, as Aldrovandus himself informs us. We exclude also the fourth and twenty-eighth species of Brisson; since the first has its bill hooked, and its thighs clothed with feathers to the knee, and the second has a short bill, which rather belongs to the Crane, Lastly, the ninth species of the same author is only the female of the Night Heron.

The CRAB-CATCHERS.

Les Crabiers. Buff.

THESE are ſpecies of Herons ſtill ſmaller than the European Egrets: they feed on lobſters and crabs; whence their name. Dampier and Wafer found them at Braſil, and at Timor in New Holland. They are therefore ſpread through both hemiſpheres. Barrere ſays, that though thoſe of America catch crabs, they alſo eat fiſh, and haunt the margins of freſh water, like the Herons. We know of nine ſpecies in the Old Continent, and of thirteen in the New.

CRAB-CATCHERS of the Old Continent.

The SQUAIOTTA HERON.

Le Crabier Caiot. Buff.

First Species.

Ardea Squaiotta. Gmel.
Cancrofagus. Briff.
Squaiotta. Aldrov. Will. Johnft. Charleton and Ray.

ALDROVANDUS fays, that this bird is called *Squaiotta* by the people of Bologna; probably from the refemblance of that word to its cry. Its bill is yellow and its legs green: it has a fine creft on the head, compofed of unwebbed feathers, white in the middle, and black on both edges: the upper part of the body is clothed with a frieze, rifing from the long thin pendant feathers, which form a kind of fecond mantle in moft of the Crab-catchers; in this fpecies they are of a fine rufous colour.

The RUFOUS HERON.

Le Crabier Roux. Buff.

Second Species.

Ardea Badiâ. Gmel.
Cancrofagus Caſtaneus. Briſſ.
The Cheſnut Heron. Lath.

ACCORDING to Schwenckfeld this bird is red (*ardea rubra*), that is a bright rufous, not cheſnut, as Briſſon tranſlates it. It is of the bulk of a Crow; its back is rufous, its belly whitiſh; its wings have a bluiſh tint, and their great quills are black. It is well known in Sileſia, where it is called the Red Heron (*Rodter-reger*): it neſtles in large trees.

The CHESNUT HERON.

Le Crabier Marron. Buff.

Third Species.

Ardea Erythropus. Gmel.
Cancrofagus Rufus. Briff.
Ardea Hæmatopus, ſeu Cirris. Aldrov. Will. and Ray.
The Red-legged Heron. Lath.

ITS colours are deeper beneath, and lighter on the back and the wings; the long narrow feathers which cover the head and float on the neck, are variegated with yellow and black; a red circle ſurrounds the eye, which is yellow; the bill is black at the point, and bluiſh green near the head; the legs are deep red. This bird is reckoned by Aldrovandus almoſt the leaſt of all the Herons. The ſame naturaliſt gives, as a variety, the one which Briſſon makes his thirty ſixth ſpecies; its legs are yellow, and there are a few more ſpots on the neck, than in the other: in all other reſpects, they are exactly ſimilar. We will not heſitate therefore to range them together. But Aldrovandus ſeems to have little reaſon in appropriating the name *Cirris* to this ſpecies. Scaliger proves indeed that Virgil did not mean by that term the Creſted-lark

as

as ufually tranflated, but a Shore-bird which is the prey of the Sea eagle. We cannot however infer that the *Cirris* is a fpecies of Heron; far lefs affign the Chefnut Crab-catcher in particular. [A]

[A] Specific character of the *Ardea Erythropus*: "Its head is crefted, its body is faffron inclining to bay."

The SGUACCO HERON.

Le Guacco. Buff.

Fourth Species.

Ardea Comata. Gmel.
Cancrofagus Luteus. Briff.
Sguacco. Aldrov. Will. and Ray.

THIS is alfo a fmall Crab-catcher, known in Italy in the vallies of Bologna by the name of *Sguacco.* Its back is of a ferruginous yellow; the feathers of the thighs are yellow; thofe of the belly whitifh; the feathers thin and pendant from the head and neck, are variegated with yellow, with white, and with black. This bird is bolder and more intrepid than the other Herons; its legs are greenifh, the iris yellow, encircled with a black ring.

The MAHON CRAB-CATCHER.

Le Crabier de Mahon. Buff.

Fifth Species.

Ardea Comata. Gmel.

It is ſmall, being not eighteen inches long, its wings are white; its back ruſty; the upper ſide of the neck yellowiſh rufous, and the fore part white gray; on the head is a fine long creſt of white gray and ruſty feathers.

The COROMANDEL CRAB-CATCHER.

Le Crabier de Coromandel. Buff.

Sixth Species.

This bird reſembles the preceding: its back is ſtained with the ſame rufous, its head and the lower part of the foreſide of the neck with gold and yellow rufous; the reſt of the plumage is white: but it has no creſt, which defect may be attri-

buted

buted to its ſex; and we ſhould therefore range it with the foregoing, were it not near three inches longer.

The WHITE AND BROWN CRAB-CATCHER.

Seventh Species.

Ardea Malaccenſis. Gmel.
The Malacca Heron. Lath.

THE back is brown or umber, all the neck and the head marked with long ſtreaks of that colour on a yellowiſh ground; the wing and the upper ſide of the body, white: ſuch was the plumage of this bird, which we received from Malacca; it was nineteen inches long.

The BLACK CRAB-CATCHER.

Eighth Species.

Ardea Novæ Guinea. Gmel.
The New Guinea Heron. Lath.

SONNERAT found this bird in New Guinea: it is entirely black, and ten inches long. Dampier mentions his having

ing ſeen in that country *Small Crab-catchers with a milk white plumage.* We are unacquainted with them.

The LITTLE CRAB-CATCHER.

Ninth Species.

Ardea Philippenſis. Gmel.
Cancroſagus Philippenſis. Briſſ.
* *The Philippine Heron.* Lath.

IT is the leaſt of all the Crab-catchers, and even more ſo than the *Blongios*; being only eleven inches long. It is a native of the Philippines; the upper ſide of the head, of the neck, and of the back, is of a brown rufous; and that colour is traced on the back in ſmall croſs lines, that wave on a brown ground; the upper ſurface of the wing is blackiſh, fringed with ſmall unequal feſtoons of a ruſty white; the quills of the wing and of the tail are black.

The BLONGIOS.

Tenth Species.

Ardea Minuta. Linn. and Gmel.
Ardeola. Briſſ.
The Little Bittern. Penn. and Lath.

THE Blongios ſtands at the bottom of the extenſive ſcale of Herons: it differs from the Crab-catchers, only becauſe its legs are rather lower, and its neck proportionally longer: inſomuch, that the Barbs, according to Dr. Shaw, call it *Boo onk*, long neck, or literally *Father-neck.* It extends the neck, or darts it forward as if by a ſpring, when it walks, or ſeeks its food. The upperſide of the head and of the back is black, with greeniſh reflections, and alſo the quills of the wings and of the tail; the neck, the belly, the upper ſurface of the wings, are of a tawny cheſnut, mixed with white and yellowiſh; the bill and legs are greeniſh.

It appears that the Blongios occurs frequently in Switzerland. It is ſcarcely known in the provinces of France, where it is never found, unleſs ſtrayed from its companions, driven by a guſt of wind, or purſued

pursued by some bird of prey *. It is met with on the coasts of the Levant, as well as on those of Barbary. Edwards figures one which was sent from Aleppo †: it differed from that just described, in having its colours more dilute, and the feathers on its back fringed with rusty, and those on the foreside of the neck and body, marked with small brown streaks; differences which seem to result from age or sex. Thus the Blongios of the Levant ‡, that of Barbary, and that of Switzerland, are all the same.

All the preceding species of Crab-catchers belong to the Old Continent: we now proceed to enumerate those of the New, observing the same distribution as in the Herons.

* "I saw one of these little Herons, of the size of a Blackbird, it suffered itself to be caught with the hand in the garden of the *Dames du Bon-pasteur* at Dijon: I saw it shut in a cage for breeding Canaries; its plumage resembled that of a meadow Rail; it was very lively, and continually bustled about in its cage, rather from a sort of inquietude, than from a desire to escape; for when a person approached, it stopt, threatened with its bill, and darted like a spring. I never met with this very small Heron in any of the provinces where I have fowled; it must be a bird of passage." *Note communicated by M. Hebert.*

† *The Little Brown Bittern.* Edwards, *Gleanings.*

‡ *Ardeola Nævia* of Brisson. (Mr. Latham supposes it to be the female. T.)

[A] Specific character of the *Ardea Minuta*: "Its head is smooth, its body brown, below tawny, its tail-quills black greenish, its stripes yellowish."

CRAB CATCHERS of the New World.

The BLUE CRAB-CATCHERS.

First Species.

Ardea Cærulea. Linn. and Gmel.
Ardea Cyanea. Klein.
Ardea Plumbea. Brown.
The Blue Heron. Catesby. Penn. and Lath.

WHAT is very singular in this bird, its bill is blue like the rest of the plumage; and were not the legs green, it would be entirely blue The feathers of the neck and head have a fine violet gloss; those of the lower part of the neck, behind the head, and on the lower part of the back, are thin and pendant; the last are a foot long, cover the tail, and project four fingers beyond it. The bird is rather smaller than a Crow, and weighs fifteen ounces. It is seen in Carolina, but only in the spring; and Catesby believes that it does not breed in that province, though he is ignorant whence it comes. The same beautiful species

cies is found in Jamaica, and appears to be divided into two varieties in that iſland.

[A] Specific character of the *Ardea Cærulea*: "Its head is creſted, its body blue."

The BROWN-NECKED BLUE CRAB CATCHER.

Second Species.

Ardea Cæruleſcens. Gmel.

ALL the body of this bird is of a dull blue; and though that tint is deep, we ſhould have reckoned it the ſame ſpecies with the preceding, if the head and neck had not been tawny-brown, and the bill intenſe yellow. It occurs in Cayenne, and may be about nineteen inches long.

The IRON-GRAY CRAB-CATCHER.

Third Species.

Ardea Violacea. Linn. and Gmel.
Ardea Stellaris Cristata Americana. Klein.
Cancrofagus Bahamensis. Briss.
Ardea Cæruleo-Nigra. Ray and Sloane.
The Gray-crested Gaulding. Brown.
The Crested Bittern. Catesby.
The Yellow Crowned Heron. Penn. and Lath.

ALL the plumage is of a dull, blackish blue, except the upper side of the head, which is ornamented with a pale, yellow crest, from which three or four white feathers rise on the back of the head. There is also a broad white stripe on the cheek, reaching to the corners of the bill: the eye is protuberant, the iris red, and the eye lid green; long slender feathers grow on the sides of the back, and fall over the tail; the thighs are yellow, the bill is black and strong, and the bird weighs a pound and an half. It is seen, says Catesby, in Carolina, during the rainy season; but is more numerous in the Bahama Islands, and breeds in the bushes that grow out of the clifts of the rocks; and so great is is the multitude in ſome of the iſlands, that, in a few hours, two men may load a canoe with the young; for though able to fly, their motions are laborious, and they ſuffer themſelves to be caught through mere ſupineneſs. They live more on crabs than on fiſh, and the ſettlers call them *crab-catchers*. Their fleſh, ſays Cateſby, is very well taſted.

The RED-BILLED WHITE CRAB-CATCHER.

Fourth Species.

Ardea Æquinoctialis. Linn. and Gmel.
Ardea-Carolinenſis Candida. Briſſ. and Klein.
The Red billed Heron. Penn.
The Little White Heron. Cateſby and Penn.

A RED bill, green legs, the iris yellow, and the ſkin that encircles it red, are the only colours which interrupt the fine white plumage of this bird. It is ſmaller than a Crow: it is found in Carolina in the ſpring, but never in winter: its bill is a little curved, and Klein remarks that many of the foreign ſpecies of Herons have not ſuch ſtraight bills as thoſe of Europe.

[A] Specific character of the *Ardea Æquinoctialis*: "Its head is ſmooth, its body white, its two firſt wing-quills brown on their outer margin."

The CINEREOUS CRAB-CATCHER.

Fifth Species.

Ardea Cyanopus. Gmel.
Ardea Americana Cinerea. Briſſ.

THIS bird inhabits New Spain, and is not larger than a Pigeon. The upper ſurface of the body is light aſh-colour; the quills of the wing are partly black, partly white; the bill and legs are bluiſh. From theſe colours we may perceive that Father Feuillée is miſtaken in claſſing it with the Bitterns.

The PURPLE CRAB-CATCHER.

Sixth Species,

Ardea Spadicea. Gmel.
Ardea Mexicana Purpuraſcens. Briſſ.
The Mexican Heron. Lath.

SEBA ſays that this bird was ſent to him from Mexico; but he applies to it the name *Xoxouquihoactli*, which Fernandez beſtowed on a ſpecies twice as large.

It is only a foot long; the upper ſide of the neck, of the back and of the ſhoulders, is purple cheſnut; the ſame tint diluted, covers all the underſide of the body; the quills of the wing are deep bay; the head is light bay, and its top black.

The CRACRA.

Seventh Species.

Ardea-Cracra. Gmel.
Cancrofagus Americanus. Briſſ.

CRACRA is the cry of this bird on the wing, and the name which the French ſettlers at Martinico have given to it. The American natives call it *Jaboutra*; and Father Feuillée, who found it in Chili, deſcribes in the following terms: it is as large as a *well grown Hen*, and its plumage is much variegated; the crown of the head is aſh-blue; the top of its back, tawny, mixed with the colour of dry-leaves; the reſt of the upper ſurface is an agreeable mixture of aſh-blue, of brown-green, and of yellow; the coverts of the wing are partly of a dull green, edged with yellowiſh,

lowiſh, and partly black; the quills are of this laſt colour, and fringed with white; the throat and breaſt are variegated with ſpots of filemot, on a white ground; the legs are of a fine yellow.

The CHALYBEATE CRAB-CATCHER.

Eighth Species.

Ardeola. Marcg. Johnſt. and Will.
Cancrofagus Braſilienſis, Briſſ.
The Blue Heron. Lath.

THE back and the head of this bird are *chalybeate* or ſteel-coloured; it has long greeniſh quills in the wing, marked with a white ſpot at the tip; the upper ſide of the wing is variegated with brown, yellowiſh, and ſteel-colour; the breaſt and belly are white, variegated with cinereous and with yellowiſh. The bird is of the ſize of a Pigeon; it occurs in Brazil, and this is all that we learn from Marcgrave.

The GREEN CRAB-CATCHER.

Ninth Species.

Ardea Virescens Linn. and Gmel.
Ardea Stellaris Minima: Klein.
Cancrofagus Viridis. Briss.
The Small Bittern. Catesby.
The Green Heron. Lath.

THIS bird is very rich in its colours, and one of the most beautiful of the genus. Long feathers of gold-green cover the upper side of the head, and form into a crest; feathers of the same colour, narrow, and flowing, cover the back; those of the neck and breast are rufous or deep reddish: the great quills of the wing are very dull green; the coverts of the wing, bright gold green, and most of them edged with fulvous or chesnut. This handsome bird is seventeen or eighteen inches long; it feeds on frogs and small fish, as well as on crabs. It appears in Carolina and Virginia only in summer; and it probably retires in autumn to warmer climates, to pass the winter.

The SPOTTED GREEN CRAB-CATCHER.

Tenth Species.

Ardea Virefcens. Var. 1. Gmel.
Cancrofagus Viridis Nævius. Briff.

THIS is rather fmaller than the preceding, but differs little in its colours; only the feathers on its head and on the nape are of a dull gold green, gloffed with bronze, and the long flender feathers on the mantle are alfo gold green, though lighter; the quills of the wing are deep brown, and their outer edge fhaded with gold green, and thofe next the body have a white fpot at the tip; the upper furface of the wing is fprinkled with white fpots; on a brown ground, fhaded with gold green; the throat is fpotted with brown on white, the neck is chefnut, and clothed below with gray pendant feathers. This fpecies is found in Martinico.

The ZILATAT.

Eleventh Species.

Ardea Mexicana Candida. Briſſ.

WE have formed this name from the Mexican *Hoitzilaztatl.* It is entirely white, with its bill reddiſh near the point, the legs of the ſame colour; it is one of the ſmalleſt of all the Crab-catchers, being ſcarce equal in ſize to a Pigeon.

The RUFOUS CRAB-CATCHER, with Green Head and Tail.

Twelfth Species.

Ardea Ludoviciana. Gmel.
The Louiſiane Heron. Lath.

IT is ſcarce ſixteen inches long; the upper ſide of the head and tail is of a dull green; the ſame colour appears on a part of the coverts of the wing, which are fringed with fulvous; the long thin feathers on the back are tinged with faint purple; the neck is rufous as well as the belly, whoſe tint borders upon brown. This ſpecies was ſent to us from Louiſiana.

The GRAY CRAB-CATCHER, with Green Head and Tail.

Thirteenth Species.

Ardea Virefcens. Var. 2. Gmel.

THIS was fent to us from Cayenne: and it refembles the preceding in many refpects, and both of them are much related to the tenth fpecies. The head and tail are equally of a dull green, and alfo a part of the coverts of the wing; a light flate gray predominates in the reft of the plumage.

The OPEN BILL.

Le Bec-Ouvert. Buff.
Ardea Pondiceriana. Gmel.
The Pondicherry Heron. Lath.

AFTER the enumeration of all the Herons and Crab-catchers, we ſhall range a bird, which, though it does not belong to their family, is more nearly related to them than to any other. It is not the buſineſs of the naturaliſt to follow the reſtraints of ſcholaſtic forms; he ſhould endeavour to trace the productions of the univerſe through their various ſhades and gradations; and the delicate tranſitions of nature are the moſt intereſting ſubjects to the eyes of a philoſopher. Such is the bird which we denominate *Open bill:* in ſome reſpects, it reſembles the Herons; and in others, it differs from them. It has, beſides, one of thoſe defects or natural imperfections, which we have remarked in a few ſpecies: its bill is wide open two-thirds of its length, both the upper and under mandible parting at that ſpace and meeting again at the point. This bird is found in India, and we received it from Pondicherry;

cherry; it has the legs and thighs of the Heron, but has only half the character noticed in the nail of the middle-toe, which ſpreads indeed into a thin plate, though it is not indented at the edges; the quills of the wing are black; all the reſt of the plumage is a light aſh-gray; the bill blackiſh at the root, the reſt of it white or yellowiſh, and thicker and broader than that of the Heron. Total length of the bird thirteen or fourteen inches. We are not informed of its natural habits. [A]

[A] Specific character of the *Ardea Pondiceriana*: "It is dirty aſh-colour, its wing-quills long and black, its middle nail not ſerrated."

The BITTERN.*

Le Butor. Buff.

Ardea Stellaris. Linn. and Gmel.

Botaurus. Briff.

The Myredromble. Turner †.

The Bittour, Bittern, or Mire-drum. Will.

THOUGH the Bitterns refemble much the Herons, their differences are fo marked as eafily to difcriminate them. Their legs are longer than thofe of the Herons, their body rather more flefhy, and their neck thicker clothed with feathers, which make it look larger. Notwithftanding the difgraceful implication of its name ‡, the Bittern is not fo ftupid as the Heron, though it is more favage: it is hardly ever feen; it inhabits only marfhes of a certain limited extent, and abounding with rufhes;

* In Greek Αστεριας, Ερωδιος Αστεριας, Οκνος: in Latin *Ardea Stellaris, Butio,* in Italian, *Tarabufo, Trombotto, Trombone* or trumpeter: in Portuguefe, *Gazola:* in German *Meer-Rind* (Sea-ox), *Mofs-Ochs* (Mofs-ox), *Rohr-Trummel* (Reed-drum), *Rofs-Reigel* (Horfe-heron), *Waffer-Ochs* (Water-ox), *Erd Bull* (Earth-bull); names that allude to the bellowing noife which this bird makes in the marfhes: in Dutch *Pitoor:* in Swedifh and Danifh *Roer-Drum* (Reed-drum): in Polifh *Bak* or *Bunk*: in Turkifh *Gelve.*

† That is, *the Mire-drum,* from the German *Trummel.*

‡ *Butor,* in French, fignifies alfo *lubber.*

it

No. 182

THE BITTERN.

it prefers the large pools ſkirted with wood: there, it leads a lonely peaceful life, covered with reeds, ſheltered from the wind and rain, and concealed equally from the hunter whom it dreads, and from its prey, which it watches. It continues whole days in the ſame ſpot, and ſeems to place its ſafety in concealment and inaction. The Heron is more reſtleſs, and ventures abroad every evening; at which time the fowlers expect it at the edge of the reedy-fens, where it alights: the Bittern, on the contrary, riſes in the duſk, and takes a final departure; and thus theſe two birds, though they inhabit the ſame tracts, never aſſociate together,

It is in autumn only and at ſun-ſet, according to Willughby, that the Bittern commences its journey, or changes its abode. In its flight it might be taken for a Heron, did it not utter from time to time a quite different cry, deeper and more reſounding, cōb, cōb. But this is ſtill leſs diſagreeable, than the frightful voice, to which the bird owes its name *: it is a ſort of lowing hi-rhōnd, repeated five or ſix times in ſucceſſion in the ſpring, and which may be heard at the diſtance of

* *Botaurus*, from *boatus* and *taurus*, denotes the *bellowing of a bull*. *Willughby*.

half

half a league. This *bumping*, as it is called, exceeds the grunt of the largeſt baſs ſtring. Could we imagine this alarming ſound to be the expreſſion of tender love? It is indeed only the call of a rude and wild bird, ſtimulated by luſt, not mollified by attachment: and as ſoon as his appetite is ſatisfied, he deſerts or rejects the female, though ſhe plies him with her aſſiduous careſſes †; nor are her ſolicitations ſufficient to incite him to repeat the almoſt momentary embrace. Accordingly the cock and hen live ſeparate, "I have often, ſays M. Hebert, put up two of theſe birds at the ſame time, and I conſtantly perceived that they ſprang more than two hundred paces from one another, and alighted at an equal diſtance." But the moments of fruition muſt return perhaps after long intervals, if it be true that the Bittern is in ſeaſon during the whole time that he *bumps* ‡; for the lowing begins in Febru-

† According to Salerne, ſuch is the indolence of the male, that the female alone takes all the trouble of courtſhip, and of the rearing of the young, "It is ſhe that ſolicits and invites him to love by the frequent viſits which ſhe pays him, and by the abundance of food which ſhe brings." But all theſe particulars, taken out of an old moral diſcourſe, (*Diſcours de M. de la Chambre, ſur l' amitié*), are probably romance.

‡ Willughby.

ary

ary §, and is yet heard in the harveſt. The people of the country ſay, that to make this noiſe the Bittern plunges its bill in the mud ||: the firſt note reſembles indeed a ſtrong inſpiration, and the ſecond an expiration rebellowed in a cavity *. But it would be difficult to aſcertain the truth of this aſſertion; ſince the bird lurks always ſo cloſe as to eſcape ſight, and the fowlers cannot reach the ſpots where it ſprings, without wading through the reeds in water often as high as the knee.

All theſe precautions for concealment and protection are fortified by caution and artifice.

§ It is certainly the cry of the Bittern which Ariſtotle conſiders in his problems (*Sect. ii.* 35), where he ſpeaks of a bellowing like that of a bull, which is heard in the ſpring from the heart of the marſhes, and of which he ſeeks a phyſical explanation in the winds impriſoned under the water and burſting from their caverns.

|| This notion is well expreſſed by a charming poet:—

——————"So that ſcarce
The Bittern knows his time, with bill ingulpht,
To ſhake the ſounding marſh."

THOMSON.

* Aldrovandus has inquired into the ſtructure of the *trachea-arteria* with a reference to the production of this ſound; many ſtrong voiced Aquatic birds, ſuch as the Swan, have a double *larynx*; the Bittern, on the contrary, has none, but the trachea, where it forks, forms two inflated ſacs, of which the rings of the trachea cover only one ſide; the other is covered with a thin ſkin, expanſible, and elaſtic; it is from theſe inflated ſacs that the impriſoned air eſcapes with violence in bellowing.

artifice. The Bittern ſits with its head erect, which being more than two feet high, it eaſily ſees over the reeds, without being perceived by the ſportſman. It does not change its ſeat until the approach of night in the autumnal ſeaſon, and paſſes the reſt of its life in ſuch inactivity, that Ariſtotle gives it the epithet *lazy* *. Its whole motion conſiſts in ſeizing a frog, or ſmall fiſh which throws itſelf in the way of the indolent catcher.

The appellation of ſtary, αστεριας or *ſtellaris*, given by the ancients to the Bittern, is derived, according to Scaliger, from its evening flight, when it ſoars aloft, and ſeems to loſe itſelf in the ſtarry-vault: others aſſert that the name refers to the ſpots ſcattered on its plumage. But theſe are diſpoſed more like daſhes than ſtars: the whole body is covered with blackiſh ſpeckles: they are ſtrewed tranſverſely on the back upon a fulvous brown ground, and run longitudinally on a whitiſh ground upon the foreſide of the neck, upon the breaſt and the belly: the bill has the ſame ſhape as in the Heron; both it and the legs are greeniſh, its opening is very wide, and the cleft extends beyond the eyes, ſo that they

* Οκνος, Hiſt. An. l. ix. c. 18.

may

may be ſaid to be ſituated upon the upper mandible: the hole of the ear is large; the tongue ſhort and ſharp, and does not reach half the bill; but the throat may be opened ſo wide as to admit the fiſt *: its long nails claſp the reeds, and ſupport it upon their floating wrecks †: it catches many frogs; and in autumn, it goes to the woods in purſuit of mice, which it ſeizes with great dexterity and ſwallows entire ‡: and in that ſeaſon, it grows very fat §. When caught it ſhews much rancour ||, and ſtrikes chiefly at the eyes **. Its fleſh muſt be of a bad quality, though it was formerly eaten, when that of the Heron was held in eſtimation ††.

The Bittern lays four or five greeniſh light-brown eggs ‡‡; and makes her neſt amidſt the reeds, or a bunch of ruſhes. Belon ſurely confounded it with the Heron, when he aſſerted, that it breeds on the tops

* Willughby.

† The great length of the nails, particularly of the hind one, is remarkable; Aldrovandus ſays, that in his time it was uſed as a tooth-pick

‡ Willughby.

§ Schwenckfeld.

|| *Idem.*

** Belon.

†† *Idem.*

‡‡ Mr. Pennant ſays five or ſix. T.

of

of trees *. That naturaliſt ſeems alſo to miſtake it for the *Onocrotalus* of Pliny, though the Roman delineates it by diſcriminating characters. There is another paſſage of Pliny, which, according to Belon, refers to the Bittern: "In the diſtrict of Arles, there is a bird which imitates the lowing of oxen, and is called the *Bull* (*Taurus*) though it is little †." In that caſe the epithet *little* is applied to the Bittern only by way of contraſt to the appellation, *Bull*.

The Bittern occurs wherever there are marſhes of ſufficient extent: it is known in moſt of the French provinces; it is not uncommon in England ‡; it is frequent in Switzerland §, and in Auſtria ||; it is ſeen alſo in Sileſia **, Denmark ††, and in Sweden ‡‡. The moſt northern parts of America have alſo their ſpecies of Bittern; and other ſpecies occur in the ſouthern regions. It appears however, that our Bittern is not ſo hardy as the Heron, and cannot ſupport our winters, but removes

* Geſner was not better acquainted with its neſt, when he ſaid that it layed twelve eggs.

† *Lib. x.* 57.

‡ Britiſh Zoology.

§ Geſner.

|| *Elench. Auſtr.* 348.

** Schwenckfeld.

†† Brunnich, *Ornithol. Boreal.*

‡‡ *Fauna Suecica.*

when

when the cold becomes exceſſive. Intelligent ſportſmen aſſure us, that they never found it by the ſides of rivulets or near ſprings in the inclement ſeaſons: and Willughby ſeems to regard its lofty flight after ſun ſet in autumn, as its departure to warmer climates.

No obſerver has given us fuller information on the ſubject of this bird than M. Baillon; and I ſhall here give an extract of the account which he obligingly ſent to me.

"The Bitterns are found almoſt every ſeaſon of the year at Montreuil-ſur-Mer, and on the coaſts of Picardy, though they are migratory. They are ſeen in great numbers in the month of December, and ſometimes dozens lurk in a ſingle tuft of reeds.

"Few birds make ſo cool a defence; it never attacks, but if once aſſailed, it fights with int epidity and temper. If darted upon by a bird of prey, it does not fly; it ſtands erect and receives the ſhock on the point of its bill, which is very ſharp; and its wounded antagoniſt retreats ſcreaming. Old Buzzards never attack the Bittern, and the Common Falcons never ſeize it, but by ruſhing upon it behind, while it is on

 the

the wing. It even makes resistance when wounded by the sportsman; and instead of retiring, it waits his onset, and gives such vigorous pushes with its bill as to pierce the leg through the boots; and many have in this way been sorely wounded. It must be killed by blows, for it would contend till death.

"Sometimes, though seldom, the Bittern turns on its back like the rapacious birds, and fights both with its bill and its claws, which are very long. It throws itself into that posture, when surprized by a dog.

The patience of this bird is equal to its courage; it remains whole hours, without stirring, in the water concealed by the rushes: it watches the eels and frogs; and except in the love season, when it takes some excercise, it is as indolent and melancholy as the Stork. At other times, it cannot be discovered but by dogs. During the months of February and March, the males utter, in the morning and evening, a cry, which may be compared to the explosion of a large musket. The females run to the sound, sometimes a dozen round one male; for the Bitterns are, like the Ducks, polygamous: the males strut among their mates, and drive off their rivals. They make

make their neſts almoſt cloſe on the water amidſt ruſhes in the month of April: the incubation laſts twenty-four or twenty-five days: the young ones are hatched naked, and of an unſightly figure, for they ſeem to be all neck and legs. They do not venture abroad until twenty days after their birth. The parents feed them firſt with leeches, lizards, and frogs'-ſpawn, and afterwards with ſmall eels. Their feathers are rufous at firſt, as in the adults; their bill and legs rather white than green. The Buzzards, which plunder the neſts of moſt of the Marſh-birds, ſeldom touch thoſe of the Bittern. The parents maintain a conſtant guard and defence. Children dare not approach, for they ſhould riſk the loſs of their eyes."

"It is eaſy to diſtinguiſh the males by their colour and their ſize, they being more beautiful and larger than the females, and their plumage having more the rufous tint; the feathers too on the breaſt and neck are longer.

"The fleſh of this bird, particularly that of the wings and breaſt, is tolerable food, provided that the ſkin be removed, whoſe capillary veſſels are filled with an acrid oil, that ſpreads through the ſubſtance in cooking, and communicates a rank unpleasant taste." [A]

[A] Specific charcter of the Bittern, *Ardea Stellaris:* " Its head is smoothish; above, brick-coloured with cross spots; below, paler, with oblong brown spots." This bird is found also over the whole extent of North America: in Hudson's Bay it appears in May, and takes up its abode among the swamps and willows; it there lays two eggs: it is exceeding lazy, and when disturbed it flies only to a short distance.—The flesh of the Bittern tastes somewhat like hare, and is not unpleasant.

BIRDS of the Old Continent which are related to the BITTERN.

The GREATER BITTERN.

Le Grand Butor. Buff.

First Species.

Ardea-Botaurus. Gmel.
Botaurus Major. Briss.
The Greater Speckled or Red Heron. Will.

GESNER is the first who has mentioned this bird, which appears to form the shade between the Herons and the Bitterns. The inhabitants of the shores of Lago Maggiore in Italy, call it *Ruffey*, according to Aldrovandus. The neck is rufous, with spots of white and black; the back and the wings are brown, and the belly is rufous; its length from the point of the bill to the end of the tail, is at least three feet and an half, and to the nails, more than four feet; the bill is eight inches, and is yellow, as well as the legs. Aldrovandus's figure represents a crest, which is not mentioned by Gesner. But he says that its neck is slender, which seems to shew that it is not a genuine Bittern: Aldrovandus accordingly observes, that this species appears to have the characters both of the Common Heron and of the Bittern; resembling the former in the head, the spots on the breast, the colour of the back and of the wings, and in its bulk; at the same time that it is similar to the latter in the form of its legs, and in the rest of its plumage, except that it is not spotted.

The LITTLE BITTERN.

Second Species.

Ardea-Marsigli. Gmel.
Botaurus Minor. Briss.
Ardea Viride-flavescens. Klein
The Swabian Bittern. Lath.

THIS small species of Bittern was seen on the Danube by Count Marsigli. Its plumage is rusty, striped with little brown lines; the foreside of its neck is white, and its tail whitish; its bill is not three inches long. If we judge from this and its other dimensions, and admit that the

the ſame proportions obtain, we ſhall be convinced that this Bittern is the ſmalleſt of all thoſe of our Continent.

The RAYED BROWN BITTERN.

Third Species.

Ardea Danubialis. Gmel.

Botaurus Striatus.

THIS is alſo a bird of the Danube. Marſigli terms it the *Brown Bittern*, and reckons it a diſtinct ſpecies. It is as ſmall as the preceding; all its plumage is ſtriped with brown black, and ruſty lines, confuſedly intermixed, ſo that a brown colour on the whole is produced. [A]

[A] Specific character of the *Ardea Danubialis*: "It is brown, marked with lines of black and tawny; its head, ſmooth; its ſtraps, naked and yellow; its throat and breaſt, whitiſh."

The RUFOUS BITTERN.

Fourth Species.

Ardea Solonienſis. Gmel.
Botaurus Rufus. Briſſ.

ALL the plumage is of an uniform colour, light ruſty under the body, and of a deeper caſt on the back; the legs are brown, and the bill yellowiſh. Aldrovandus ſays that this ſpecies was ſent to him from Epidaurus; and he claſſes it with that of a young bird, caught in the fens near Bologna, which had not yet acquired its full colours; he adds that it ſeemed to be more related to the Bitterns than to the Herons. And perhaps, as Salerne conjectures, it may be the ſame with the ſmall Bittern, which appears ſometimes in Sologne, and is known under the name of *Quoimeau.* Marſigli aſſigns the Danube alſo as the haunt of this ſpecies, which is the third of Aldrovandus; and the authors of the Italian Ornithology ſay, that it is a native of the country of Bologna *.

It is alſo found in Alſace; for Dr. Hermann writes us, that he had one of the

* *Sgarza ſtellare roſſiccia.* Gerini, *tom. iv. p.* 50.

birds,

birds, which refuſed all ſuſtenance, and died of hunger. He adds, that notwithſtanding its long legs, this Bittern climbed up a ſmall tree, of which it could embrace the trunk, while its bill and neck were vertical and in the ſame line †. [A]

† Extract of a letter from Dr. Hermann, dated Straſburg, 22d September 1779.

[A] Specific character of the *Ardea Solonienſis*: "Its top is black; its ſmooth head and its neck are ferruginous; its body is blackiſh above, and tawny below."

The LITTLE SENEGAL BITTERN.

Fifth Species.

Ardea Senegalenſis. Gmel.
The Senegal Bittern. Lath.

THE ſhort, thick clothed neck of this bird belongs more to the Bitterns than to the Herons; and we therefore range it among the former. It is very ſmall, not exceeding a foot in length. [B]

[B] Thus deſcribed by Mr. Latham: "It is brown; its belly, its wings, and its tail, are white; its head and neck, ſtreaked with black; on the middle of the wings there is a longitudinal bar of pale rufous." He adds, that it is twelve inches long; that its bill is brown, its legs yellow, and the feathers of its neck looſe and all rufous.

The SPOTTED BITTERN.

Sixth Species.

Ardea-Gardeni Gmel.

Botaurus Nævius. Briss.

The Spotted Heron. } Lath. and Penn.
The Gardenian Heron. }

SPORTSMEN call this bird the *Sloven* *. It is as large as a Crow, and measures more than twenty inches from the bill to the nails. All the ground of its plumage is brown, which is deep on the quills of the wing, and light on the foreside of the neck and under the body; sprinkled on the head, on the upperside of the neck, on the back, and on the shoulders, with small white spots, placed at the tips of the feathers: each of the wing-quills is terminated by a white spot.

To the same species we shall refer the *Cayenne Pouacre*, No. 939: Pl. Enl. the only difference being, that the plumage on the back has a blacker ground, and that the foreside of the body is spotted with brown dashes on a whitish ground †.

* *Pouacre*, a lousy lubberly fellow.

† This seems to be the same with the Gardenian Heron, *Ardea Gardeni*: "It is brown, its head smooth, its back blackish; its throat and breast whitish spotted with brown." Mr. Pennant has named it after Dr. Garden, an ingenious physician, who formerly resided in South Carolina.

BIRDS of the New Continent which are related to the BITTERN.

The STARRED-BITTERN.

L' Etoilé. Buff.

First Species.

Ardea Virescens. Var. 2d. Lath.
Ardea Fusca. Klein.
Botaurus Americanus Nævius. Briss.
Ardea Stellaris Minor. Ray and Sloane.
The Crab-catcher. Brown.
The Brown Bittern. Catesby and Lath.

WE have given this bird the epithet of *starrea*, because its plumage, which is entirely brown, is sprinkled on the wing with a few white spots thrown irregularly, and that give it some resemblance to the preceding species. It is somewhat smaller than the European Bittern; it haunts pools and rivers remote from the sea and in the highest parts of the country. Besides this species, which is spread through most of the provinces of North America, there seems

ſeems to be another in Louiſiana more like the European *.

* The Bitterns are Aquatic Birds that live on fiſh: they have a very large bill: they are known in France, and ſhall therefore omit ſaying any thing more of them," Le Page Dupratz, *Hiſtoire de la Louiſiane*, tom. ii. p. 218.

The YELLOW BITTERN OF BRAZIL.

Second Species.

Ardea Flava. Gmel.
Botaurus Braſilienſis. Briſſ.
Ardea Braſilienſis roſtro ſerrato. Ray. and Will.

WE may conclude from the proportions aſſigned by Marcgrave, that it is a Bittern. It is as large as a Heron; its neck is a foot in length; and its body five inches and a half; its tail four; and its thigh and leg more than nine; all the back and the wing are clothed with brown feathers, waſhed with yellow; the quills of the wing are partly black, partly cinereous, and interſected tranſverſely with white lines; the long feathers, which hang from the head and neck, are of a pale yellow, waved with black; thoſe on the underſide of the neck, on on the breaſt, and on the belly, are white, waved with brown and fringed round with yellow. We ſhall remark, as a ſingular circumſtance, that the bill is indented near the point both above and below.

The LITTLE BITTERN OF CAYENNE.

Third Species.

Ardea Undulata. Gmel.
The Zigzag Bittern. Lath.

THIS Little Bittern is a foot or thirteen inches long; all its plumage has a ruſty gray ground, and is ſpotted with dark brown in ſmall croſs lines very cloſe together, waved and twiſted in zigzags, and dotted below the neck, on the ſtomach, and on the flanks; the upperſide of the head is black, the neck very thick of feathers, and appears to have as much girth as the body.

The HUDSON'S BAY BITTERN.

Fourth Species.

Ardea Stellaris. Var. Gmel.
Botaurus Freti Hudſonis. Briſſ.

THE garb common to all the Bitterns is a plumage of rufous or ruſty ground, broken more or leſs and interſected with lines, and brown or blackiſh ſtreaks; and this is the garb of the Hudſon's Bay Bittern. It is ſmaller than the European, its length from the bill to the nails being ſcarcely two feet and an half. [B]

[B] Thus deſcribed by Briſſon: " Above, it is ruſty, ſtreaked acroſs with blackiſh; below, whitiſh, variegated with tawny longitudinal ſpots, ſprinkled with black; its top, blackiſh; the lower part of its neck white, variegated with tawny longitudinal ſpots, margined with black; the feathers at the origin of the neck very long; the tail-quills tawny, ſtreaked tranſverſely with blackiſh; the bill blackiſh above, and at the tip, below yellow; the legs, bright yellow."

The ONORE.

Fifth Species.

Ardea Tigrina. Gmel.
The Tiger Bittern. Lath

AFTER the Bitterns of the New World we place the birds termed *Onorés* in the *Planches Enluminées*. That name is applied in Cayenne to all the ſpecies of Herons; but the birds to which we reſtrict it, are more related to the Bitterns: they have the ſame form and colours, and differ only becauſe their neck is not ſo well clothed with feathers, though it has a cloſer fur, and is not ſo ſlender, as the neck of the Herons. The preſent is almoſt as large, but not ſo thick as the European Bittern; all its plumage is agreeably marked, and widely interſected, by black croſs bars in zigzags, on a rufous ground on the upper ſide of the body, and of a light gray on the under.

The RAYED ONORE.

Sixth Species.

Ardea Lineata. Gmel.

The Lineated Bittern. Lath.

THIS ſpecies is rather larger than the preceding, being two feet and a half long; the great quills of the wings and the tail are black; all the upper ſurface is handſomely inlaid with very fine ſmall lines of rufous, yellowiſh and brown, which wave tranſverſely, and form demi-feſtoons; the upper ſide of the neck and the head are of a bright rufous, interſected alſo with ſmall brown lines; the forepart of the neck and of the body is white, marked lightly with ſome brown ſtreaks.

Theſe two ſpecies were ſent to us by M. De la Borde, King's phyſician at Cayenne. They lurk in the gullies excavated by the rills which flow into the Savannas; they haunt alſo the ſides of rivers. In droughts, they lie cloſe in the thick herbage; they are fluſhed at a great diſtance, and never two appear together. If we wound one, we muſt be cautious in approaching it; for it acts on the defenſive, and

and, drawing back its neck, it ſtrikes a violent blow with its bill, aiming at the eyes. Its habits are the ſame with thoſe of our Heron.

M. De la Borde ſaw a tame, or rather, a captive *Onoré* in a houſe. It was continually on the watch for rats, and it caught them more dextrouſly than a cat. But though kept above two years, it always remained in hiding places, and if a perſon went near its retreat, it ſought with a threatning aſpect to dart at his eyes. Both ſpecies ſeem to be ſtationary in their native tracts, and are rare.

The ONORE OF THE WOODS.

Seventh Species.

Ardea Braſilienſis. Linn. Gmel. and Briſſ.
Soco Braſilienſibus. Marcg.
Cocoi Tertius. Piſon.
The Clucking Hen. Brown and Dampier.
The Braſilian Bittern. Lath.

IT is found in Guiana and Brazil. Marcgrave ranges it under the general name *Soco*, with the Herons. But it appears to be much related to the Onorés, and conſequently to the Bitterns. The back, the

 rump,

rump, and the ſhoulders, are entirely blackiſh, dotted with yellowiſh ; and, what is uncommon, the plumage is the ſame on the breaſt, the belly, and the ſides ; the upper ſurface of the neck is white, mixed with longitudinal black and white ſpots. Marcgrave ſays that the neck meaſures a foot, and that the total length, from the bill to the nail, is about three feet.

No. 183

THE NIGHT HERON.

The BIHOREAU.

Ardea-Nycticorax. Linn. and Gmel.
Nycticorax. Gesner. Aldrov, Johnst. Sibb. Briss. &c.
Ardea Varia. Klein. and Schwenck.
The Night Raven. Will. and Alb.
The Night Heron. Penn. and Lath *.

MOST of the naturalists have given this bird the appellation of *Nycticorax*, or *Night-raven*, on account of a strange sort of croaking, or rather a frightful and dismal rattling, which it makes during the night †. And this is the only resemblance which it bears to the Raven; for in other respects, it is analogous to the Heron: the only difference is, that its neck is shorter and better feathered; its head larger, and its bill thicker: it is also smaller, not exceeding twenty inches in length. Its plumage is black, with a green gloss on the head and the nape of the neck; dull green on the back; pearl-gray on the wings and the tail, and white on the rest of the body. The male has on

* In Italian *Nétticorace*: in German *Nicht-Rab* (Night raven); *Bunter Reger* (Mottled Heron) or *Schild-Reger*: (Shield Heron): in Flemish *Quack*.

† *Vesperé & noctu absonâ voce molestat.* Schwenckfeld.

on the nape of the neck ſome feathers, commonly three in number, exceedingly delicate, of a ſnowy white *, and about five inches long; and of all creſt-plumes, theſe are the moſt beautiful and the moſt precious †; they drop in the ſpring, and return only once a-year. The female wants this ornament, and differs conſiderably from the male; ſo that ſome naturaliſts have miſtaken it. Briſſon has made it his ninth ſpecies of Heron. It has all the upper ſurface of a ruſty-aſh colour; daſhes of the ſame tint on the neck; and the underſide of the body is light gray.

The *Biboreau* breeds in rocks, according to Belon, who thence derives its ancient French name, *Roupeau*: but Schwenckfeld and Willughby agree, that it builds its neſt on alders near marſhes. Theſe oppoſite accounts cannot be reconciled without ſuppoſing, that theſe birds vary their habits according to circumſtances; that in the plains of Sileſia and Holland they ſettle on the aquatic trees, and that on the coaſts of Britany, where Belon ſaw them, they

* Belon.

† "They are ſold at a high price, ſays Schwenckfeld, and our young nobility are fond of wearing the plumes in their hat."

neſtle

nestle in the cliffs. It is affirmed that they lay three or four white eggs *.

The *Bihoreau* seems to be a bird of passage. Belon saw one exposed in the market, in the month of March; Schwenckfeld asserts, that it retires from Silesia in the beginning of autumn, and returns with the Storks in the spring. It frequents equally the sea-shores, or the rivers and inland marshes. It is found in France at Sologne; in Tuscany on the lakes Fucecchio and Bientine †; but the species is every where more rare than that of the Heron: it is also less dispersed, and is not spread to Sweden ‡.

As its legs are not so tall and its neck shorter than the Heron, it lives partly on the water and partly on the land; and subsists as much upon crickets and slugs as upon frogs and fish. It remains concealed during the day, and does not stir until the approach of night, when it utters its cry *ka*, *ka*, *ka*, which Willughby compares to the groans of a person reaching §.

* Willughby and Schwenckfeld.

† Italian Ornithology, *tom. iv p.* 49.

‡ We judge so, because it is not mentioned in the *Fauna Suecica*.

§ *Nycticorax, quod interdiu clamet voce absonâ, & tanquam vomiturientis.*

 The

The Night Heron has very long toes; its thighs and legs are greenish yellow; its bill is black, and the upper mandible slightly arched; its eyes are brilliant, and the iris forms a red or orange circle round the pupil. [A]

[A] Specific character of the Night-heron, *Ardea Nicticorax*: "It has a horizontal three-feathered crest on the back of its head, its back black, its belly yellowish."

The BIHOREAU OF CAYENNE.

Ardea Cayanensis. Gmel.
The Cayenne Night Heron. Lath.

IT is as large as the European Night Heron; but is in general not so thick. Its body is more slender; its legs taller; its neck, its head, and its bill, are smaller; the plumage is of a bluish-ash colour on the neck, and on the underside of the body; the upper surface is black, fringed with cinereous on each feather; the head is enveloped with black, and the crown is white; there is also a white streak under the eye. The tuft in this bird consists of five or six feathers, of which some are white, others black. [B]

[B] Specific character of the *Ardea Cayanensis*: "It is cinereous, its head black, and its top white; the crest unequal, consisting of six feathers, partly black, partly white."

The TUFTED UMBRE.

L' Ombrette. Buff.
Scopus-Umbretta. Gmel.
Scopus *. Briss.

FOR the knowledge of this bird we are indebted to Adanson, who found it at Senegal. It is rather larger than the Night Heron. It owes its name to the dun or umbre cast of its plumage. It ought to be regarded as an anomalous species in the tribe of Marsh-birds; for it belongs not exactly to any of them. It might be classed with the Herons, if its bill had not been of an entirely different form, and even peculiar to itself; being very broad and thick near the head, growing flatter at the sides as it extends; the ridge of the upper mandible is prominent the whole length, and seems to be detatched by two grooves that run on each side, which Brisson describes, by saying that the bill appears composed of several jointed pieces: the ridge is reflected at the tip of the bill, and terminates in a curved point; the whole

* From Σκια, a shadow.

length

length of the bill is three inches and three lines. The leg is four inches and a half, and the naked part of the thigh two inches. These measures were taken from one of these birds deposited in the king's cabinet. Brisson seems to make them larger. The toes are invested near the root by the beginning of a membrane, which spreads more between the outer and middle toe: the hind toe is not joined as in the Herons to the side of the heel, but to the heel itself.

The COURLIRI, OR COURLAN.

Ardea Scolopacea. Gmel.
The Scolopaceous Heron. Lath.

THIS bird is nearly of the ſame height and bulk with the Herons. Its length from the bill to the nails is two feet eight inches; the naked part of the thigh, reckoning from the leg, is ſeven inches; the bill is four, and is ſtraight almoſt its whole length, and ſlightly curved near the point; and this is the only circumſtance in which it reſembles the Curlews. On the nail of the great toe, there is a protuberant edge on the inſide, which repreſents the indented comb of the Heron. The plumage is a beautiful brown, which becomes reddiſh and coppery on the great quills of the wing and of the tail. Each feather of the neck has a white daſh on its middle. This ſpecies is new, and was ſent to us from Cayenne under the name of *Courliri*. [A]

[A] Specific character of the *Ardea Scolopacea*: "It is brown, its throat and breaſt ſtreaked with white, its chin and legs white; its tail and its wing-quills have a copper gloſs."

The SAVACOU.

Cancroma. Linn. and Gmel.

Cancroma Cochlearia. Lath.

{ *Cochlearius.*
Cochlearius Fuſcus. Briſſ.

The Boat bill. Brown and Lath *

THE *Savacou* is a native of Guiana and Brazil. It has nearly the bulk and proportions of the *Bihoreau*; and, by its general ſtructure and its mode of living, it would approach the Heron-tribe, did not its broad and remarkable flat bill ſeparate it widely, and diſcriminate it even from all the other Marſh-birds. It has been called *Spoon-bill*, from the reſemblance to two ſpoons applied to each other at the concave ſides. On the convexity of the upper mandible, there are two deep grooves, which are ſent off from the noſtrils and produced, ſo that the middle ſpace makes a high ridge terminated by a ſmall hooked point. The lower mandible, upon which the upper is fitted, may be ſaid to be the

* Called *Savacou* or *Saouacou* at Cayenne, *Rapapa* by the Garipane ſavages, *Tamatia* in Brazil.

frame,

No. 184

THE CRESTED BOAT-BILL.

frame, whereon the ſkin prolonged from the throat is extended. Both mandibles are ſharp at the edges, and conſiſt of a ſolid and very hard horn. The bill is four inches from the corners to its point, and its greateſt breadth twenty lines.

Notwithſtanding its formidable weapon, the Savacou ſeems to have a mild diſpoſition, and to lead a calm peaceful life, if we may judge from the names applied by nomenclators: the appellation *Cancrofagus* given by Barrere implies that it ſubſiſts on crabs. But on the contrary, it removes from the ſea ſhore, and haunts the deluged Savannas, or the ſides of rivers where the tide never aſcends *. There, perched on the aquatic trees, it watches the fiſh as they paſs by it, and plunges after them under the water, and again emerges with its prey †. It walks with its neck bent, its back arched, its carriage conſtrained, and its air as mournful as that of the Heron ‡. It is ſavage, and ſhuns the ſettlements § Its eyes are placed very near the root of its bill,

* Obſervations made at Cayenne by M. Sonini de Mannoncour.

† Memoirs communicated by M. de la Borde, King's phyſician at Cayenne.

‡ *Dorſo incurvato incedens, & collo incurvato.* Marcgrave.

§ M. de la Borde.

bill, and give it a wild aſpect. If it is caught, it cracks its bill; and when it is irritated, or thrown into agitation, it erects the long feathers on the top of its head.

Barrere reckons three ſpecies of the Savacou, which Briſſon reduces to two, and which may probably be comprehended under one. In fact, the gray and the brown have no remarkable difference, except that the latter is furniſhed with a long tuft, which may be the character of the male. The other, which we conceive to be the female, has a trace of the ſame character in the feathers which hang from the back of the head. The difference between the colours of their plumage may be attributed to age or ſex; eſpecially as the *Variegated Savacou* * forms the uniting ſhade. Their forms and proportions, too, are preciſely the ſame; and we are the more convinced that they conſtitute but a ſingle ſpecies, becauſe nature, though ſhe ſports with variety in the general plan of her works, leaves ſome ſolitary and inſulated productions of peculiar ſtructure on the confines of the grand diviſions. The Avoſet, the

* Brought from Cayenne by M. Sonini.

Spoonbill,

Spoonbill, the Flamingo, &c. are examples of this remark.

The Brown and Crefted Savacou, which we take to be the male, has more of the rufous-gray than of the bluifh-gray, on its mantle. The feathers of the nape are black, and form a bunch of eight inches long, falling on the back: thefe feathers are floating and fome of them eight lines broad.

The Gray Savacou, which appears to be the female, has the whole of its mantle of a bluifh light gray, with a fmall black zone on the top of the back; the underfide of the body is black, mixed with rufous; the forepart of the neck and the front, are white; the head-drefs falls back into a point, and is of a bluifh black.

In both, the throat is naked; the fkin which covers it, feems capable of a confiderable dilatation: and this is probably what Barre means by the expreffion *ingluvie extuberante*. This fkin, according to Marcgrave, is yellowifh, as well as the legs; the toes are flender, and the *phalanges* long. We may alfo remark, that the hind-toe is articulated with the fide of the heel, near the outer toe, as in the Herons. The tail is fhort, and does not project beyond the wing. The total length of the bird is about

about twenty inches. Our meaſures were taken from larger ſpecimens than that deſcribed by Briſſon, which was probably a young one. [A]

[A] Linnæus erects the Boat-bill into a diſtinct genus, under the name of *Cancroma*, which is divided into two ſpecies, the *Cochlearia* and the *Cancrophaga*, the former having a tawny belly, and the latter a whitiſh. But Mr. Latham, with more propriety, follows our author in making only one ſpecies, including two varieties.

No. 185

THE WHITE SPOON-BILL.

The WHITE SPOONBILL.

La Spatule. Buff.
Platelea Leucorodia. Linn. and Gmel.
Platea, ſive Pelicanus. Geſner. Aldrov. Will. and Sib.
Ardea Alba. Johnſt.
The Spoon-bill or Pelican. Alb. and Ray *.

THOUGH the Spoon bill is diſtinguiſhed by a ſingular figure, nomenclators have, under improper and foreign appellations, confounded it with quite different birds. It is neither a *White-Heron*, nor a *Pelican*, as ſome have repreſented it: and the name *Spatula* or *Spoon-bill*, adopted in moſt languages, ſeems to ſuit it the beſt. The bill is flat; and near its extremity, it ſpreads like a ſhovel, and terminates in two rounded plates, thrice as broad as the body of the bill itſelf. On account of this conformation, Klein has denominated the bird *Anomaloroſter*, or *Anomalous-bill*; the ſubſtance of this bill is as anomalous as its ſhape, being flexible like leather, and therefore by no means fitted for the office which Pliny and Cicero

* In Greek Λευκορωδιος: in Latin *Platea* or *Platelea*: in Hebrew *Kaath*: in Italian *Beccaroveglia*: in German *Pelecan*, or *Loeffler*: in Flemiſh *Lepelaer*: in Swediſh *Pelecan*: in Ruſſian *Calpetre*: in Poliſh *Pelican* or *Plaſkonos*.

aſſign

aſſign to it, applying inaccurately to the Spoon-bill what Ariſtotle had juſtly aſſerted of the Pelican; viz. that it darts at the Diving birds, and bites them on the head until they reſign their prey †; Scaliger, inſtead of correcting theſe miſtakes, accumulates others: he confounds the *Plat lea and Pelican*, and then adds from Suidas that the latter was the δενδροκαλαπτης, which is the Woodpecker; thus tranſporting the Spoon-bill from the margin of lakes to the heart of foreſts, and making it bore trees with a bill which is deſtined ſolely to ſearch in the mud. Such confuſion and ſuch falſe erudition merit not an examination; and inſtead of waſting our time in obſcure inveſtigation, we ſhall proceed directly to ſurvey the ſimple beauties of nature.

The Spoon-bill is entirely white; as large as a Heron; its legs not ſo tall, its neck not ſo long, but clothed with ſmall narrow feathers; thoſe below the head are

† Ariſtotle, *Hiſt. Animal.* Lib. ix. 14. — "*Legi etiam ſcriptum hic eſſe avem quamdam quæ platelea nominetur; eam ſibi cibum quærere advolantem ad eas aves quæ ſe in mari mergerent, quæ cum emerſiſſent, piſcemque cepiſſent, uſque adeo premere earum capita mordicus, dum illæ captum amitterent, quod ipſa invaderet.* Cicero, *lib. ii. de Nat. Deor.—Platea nominatur advolans ad eas quæ ſe in mari mergunt, & capita illarum morſu corripiens, donec capturam extorqueat.* Plin. *lib. x.* 56.

are long and narrow, and form a tuft which falls behind; the throat is covered, and the eyes encircled, with a naked ſkin; the legs and the bare part of the thighs, are covered with a black, hard, and callous ſkin; a portion of membrane connects the toes near their junction, and, by its production, fringes and borders them to their extremity; black croſs waves mark the yellowiſh ground of the bill, whoſe extremity is yellow, mixed ſometimes with red; a black border, which runs contiguous to a channel, forms a ſort of ledge quite round the bill, and within, there is a long groove under the upper mandible; a ſmall point bent downwards, terminates the extremity of the ſpoon, whoſe greateſt breadth is twenty-three lines, and which is furrowed internally with ſmall *ſtriæ*, that make the ſurface rougher on the inſide than on the outſide: near the head, the upper mandible is ſo broad and thick, as to occupy apparently the whole front: the two mandibles near their origin are equally beſet within near the edges, with ſmall tubercles, or furrowed prominences, which ſerve to bruiſe ſhells, or hold a ſlippery prey; for it appears that this bird feeds equally on fiſh, on cruſtaceous animals, on aquatic inſects, and on worms.

The Spoon-bill inhabits the ſea-ſhore, and ſeldom occurs in the inland country *, except on ſome lakes †, and tranſiently by the ſides of rivers. It prefers the fenny coaſts; and is found in thoſe of Poitou, of Britany ‡, of Picardy, and of Holland: ſome places are even famous for multitudes of Spoon-bills; ſuch are the marſhes of Sevenhuys, near Leyden §.

Theſe birds neſtle on the tops of large trees near the ſea coaſt, and build with ſticks; they have three or four young. During the breeding ſeaſon, they are very noiſy in their retreats, and return regularly every night to repoſe ‖.

Of the four Spoon-bills deſcribed by the Academicians ¶, and which were entirely white, two had a little black on the end of the wing; which does not denote a difference of ſex, as Aldrovandus preſumes, for it occurs in both the male and the female. The tongue is very ſmall, of a triangular ſhape, and not exceeding three

* Salerne.

† As on thoſe of Bientina and Fucecchio in Tuſcany, according to Gerini: that author is miſtaken in calling this bird a *Pelican*.

‡ Belon.

§ Albin, and Johnſt.

‖ Belon.

¶ Mem. de l' Acad. depuis 1666 juſq' en 1669: *tom. iii. par.* 3. *p.* 27 & 29.

lines

lines in all: the *æsophagus* dilates as it descends; and it is probably in this cavity, that the bird detains and digests the small muscles and other shell fish, which are swallowed, and the shells rejected after the pulp is decocted and extracted *: the gizzard is lined with a callous membrane, as in the granivorous birds; but instead of the *cæcum*, which is found in these, it has only little and exceeding short protuberances at the extremity of the *ileon*; the intestines are seven feet long: the *trachea arteria* is like that of the Crane, and makes a double inflection in the thorax: the heart is furnished with a *pericardium*, though Aldrovandus says that he could not perceive it †.

These birds penetrate in summer as far as West Bothnia and Lapland, according to Linnæus: they appear also in small numbers in Prussia, during the autumnal rains, having arrived from Poland ‡. Rzaczynski says, that they are seen, though rarely, in Volhinia: some pass into Silesia, in the

* *Platea cum devoratis se implevit conchis, calore ventris coctas evomit, atque ex iis esculenta legit, testas excernens.* Plin. *Lib. x.* 56.

† Mem. de l' Acad. *uti supra.*

‡ Klein.

months of September and October *. They inhabited, as we have already noticed, the western coasts of France. They are found also on those of Africa, at Bissao near Sierra Leona †; in Egypt, according to Granger ‡; at the Cape of Good Hope, where Kolben relates, that they live both on serpents and fish, and are called *Slangen-vreeter*, or *Serpent-eaters* §: Commerson saw them in Madagascar, where the inhabitants denominate them *Fangali-am bava*, or *Spade-bills* ||. The Negroes in some parts name them *Vang-van*, and in others *Vourou-doulon*, or the Devil's-birds ¶. The species therefore though not numerous is widely diffused, and seems even to have made the circuit of the Ancient Continent. Sonnerat found Spoon-bills in the Philippine islands **;

* *Aviar. Siles. p.* 314. Schwenckfeld seems here to confound the Pelican with the Spoon-bill; for he relates at this place, from Isiodorus and St. Jerome, the fable of the resurrection of the young of the Pelican, by the blood which it discharges from its breast, when the serpent has killed them.

† Brue, *Hist. Gen. des Voy. tom. ii. p.* 590.

‡ Voyage de Granger. *Paris*, 1745. *p.* 237.

§ His account is not altogether accurate, and he improperly terms the bird *Pelecan*; but his figure is that of the Spoon-bill.

|| *Vourou-gondron*, according to Flaccourt.

¶ The Negroes give them this name, because their cry is believed to forebode the death of some person in the village. *Note left by M. Commerson.*

** Voyage a la Novelle Guinée, *p.* 89.

and

and though he diſtinguiſhes two ſpecies, the want of the creſt, which conſtitutes the chief difference, does not to us appear to make a ſpecific character; and hitherto only one ſpecies is known, which is nearly the ſame from the North to the South of our hemiſphere. It occurs alſo in the New World, and though the ſpecies has been here divided into two, we may join them together; and their reſemblance is ſo ſtrong to the European Spoon-bill, that we may impute the ſmall differences to the influence of climate. [A]

The American Spoon-bill * is only a little ſmaller in all its dimenſions than the European; it differs alſo by the roſe or carnation which paints the white ground of its plumage on the neck, the back, and the ſides; the wings are more ſtrongly coloured, and the red tint runs into a crimſon on the ſhoulders and the coverts of the tail, of which the quills are rufous; the ſhaft of thoſe of the wing is marked with fine

[A] Specific character of the White Spoon-bill, *Platalea Leucorodia*: "Its body is white, its throat black, the back of its head ſomewhat creſted.

* *Platalea-Ajai.*
Platalea-Ajai. Var. Linn. and Gmel.

Platea Roſea.
Platea Coccinea. Briſſ.

The Braſilian Roſeate Spoon-bill.
The Scarlet Spoon bill. Lath.

 carmine;

carmine; the head and the throat are naked. These beautiful colours are found only in the adult; for there are some which have much less red, and are even almost entirely white, the head not bare, and the quills of the wing partly brown; which are the vestiges of its first garb. Barrere affirms, that there is the same progress of colour in the American Spoon bills as in many other birds, the Red Curlews, for instance, and the Flamingos, which during the two first years are almost entirely gray or white, and do not become red till the third year. It hence follows that the Rose-coloured Bird of Brazil, or the *Ajaia* of Marcgrave, described in its early stage, with wings of pale carnation; and the Crimson Spoonbill of New Spain, or the *Tlauhquechul* of Fernandez, described in its adult state, are really the same. Marcgrave says, that numbers of them are seen on the river St. Francis or Seregippa, and that the flesh is pretty good. Fernandez ascribes to it the same habits with our Spoon bill; that it lives by the Sea-side on small fishes, that these must be given to it alive if it is to be reared in the domestic state *, *hav-*

* The European Spoon-biil will live in confinement: it may be fed, says Belon, with fowls' guts. Klein kept one a long time in his garden, though its wing had been broken by a shot.

ing

ing found by experience, he adds, *that they will not touch dead fish* †,

This Rose coloured Spoon bill is diffused in the New Continent, as the white one in the old, over a great extent from North to South ; from the coasts of Mexico and Florida ‡, to Guiana and Brazil It is found too in Jamaica §, and probably in the adjacent islands. But the species is no where numerous : at Cayenne, for instance, there are perhaps six times more Curlews than Spoon-bills, and their greatest flocks never exceed eight or nine at most, commonly only two or three, and often these are accompanied by Flamingoes. In the morning and evening, the Spoon bills are seen on the Sea shore, or sitting on trunks that float near the beach ; but about the middle of the day in the sultriest weather they enter the creeks, and perch very high on the aquatic trees. Yet they are

† This peculiarity perhaps has induced Nieremberg to term it *avis vivivora.*

‡ Page du Pratz, *Hist. de la Louisiane, tom. ii. p.* 116. "We have received from Balize (in New Orleans) a large bird called the *Spatula,* because its bill is of that form : its plumage is white, which turns into a light red : it grows familiar, and remains in the court-yard." *Extract of a letter from M. de Fontette, 20th October* 1750.

§ The American Scarlet Pelecan, or Spoon-bill. *Tlauhquechul* Fernand. *Ajaia* Brasil, &c. Sloane, *Jamaica, Vol. xi. p.* 317.

not very wild; for at ſea they paſs very near the canoes, and on land they ſuffer a perſon to get within gun ſhot of them, whether they be alighted or on the wing. Their beautiful plumage is often ſoiled by the mud, in which they wade deep in queſt of prey. M. De la Borde, who made theſe obſervations on their œconomy, confirms M. Barrere's account of their colour, and aſſures us that the Spoon-bills of Guiana do not aſſume, until about their third year, that beautiful red tint, and that the young ones are almoſt entirely white.

M. Baillon, to whom we are indebted for many good obſervations, admits two ſpecies of Spoon-bills, and informs me, by letter, that both theſe appear on the coaſts of Picardy, in the months of November and April, but that neither of them remains: they ſtop a day or two near the ſea, and the adjacent marſhes; they are few in number, and ſeem extremely ſhy.

The firſt is the Common Spoon bill, which is of a very bright white and has no creſt: the ſecond ſpecies is creſted, and ſmaller than the other. M. Baillon thinks that theſe differences, with ſome varieties in the colours of the bill and plumage, are ſufficient to conſtitute two diſtinct and independent ſpecies.

He

He is alſo perſuaded, that all the Spoon-bills are hatched gray like the Egrets, which they reſemble in the ſhape of their body, in their manner of flying, and in their other habits: he regards thoſe of St. Domingo as forming a third ſpecies. But it appears to us from the reaſons already advanced, that theſe are only three varieties of the ſame ſpecies, ſince they have all the ſame inſtincts and habits.

M. Baillon obſerved in five Spoon bills, which he was at the pains to diſſect, that all of them had their ſtomach filled with ſhrimps, ſmall fiſh, and water inſects; and, as their tongue is almoſt nothing, and their bill neither ſharp nor indented, it would ſeem that they cannot catch eels, or any fiſh that make reſiſtance, and that they live on very ſmall animals, which obliges them to ſearch continually for their food.

It is probable that theſe birds in certain circumſtances make the ſame clattering with their bill as the Storks; for M. Baillon, having wounded one, obſerved that it made that noiſe, which was produced by the quick and ſucceſſive motion of the mandibles, though the bill was ſo weak that it could hardly ſqueeze the finger.

[A] Specific character of the Roſeate or Scarlet Spoon-bill *Platalea-Ajai*: "Its body is blood coloured."

The WOODCOCK.*

La Becasse. Buff.
Scolopax Rusticola. Linn. and Gmel.
Scolopax. Briss. Ray. Aldrov. Johnst. &c.

Of all the birds of passage, the Woodcock is that on which sportsmen set the most value; both on account of the excellence of its flesh, and the facility with which it is caught. It arrives in our woods about the middle of October, at the same time with the Thrushes †; it then descends from the lofty mountains, which it inhabited during the summer, and seeks a milder air in the plains ‡: for it does not migrate into distant

* In Greek Σκολοπαξ, which also signifies *a stake*, and was applied to the Woodcock because of the length of its bill: in Latin *Perdix Rustica, Rusticula, Gallinago*: in Italian *Becassa, Beccacia, Gallinella, Gallina* with the epithets *arciera, rusticella* and *salvatica*; in Rome *Pizzarda*; in Tuscany *Acceggia*; and in Lombardy *Gallinacia*; in German both the Woodcock and Snipe have the general appellation *Schnepffe* modified with various epithets; those of the Woodcock *gross, wald, holtz berg* &c. (*great, wood, forest, mountain*, &c): in Flemish *Sneppe*: in Polish, *Slomka* and *Pardwa*: in Swedish, *Merkulla*: in Norwegian *Blom-Rokke, Rutte*: in Danish, *Holt-Sneppe*: in Turkish, *Tcheluk*.

† Aloysius Mundella, *apud Gesnerum*.

‡ The time of fowling for Woodcocks is well depicted by the poet Nemesianus.

Cum nemus omne suo viridi spoliatur honore
——— *præda est facilis & amœna scolopax.*

Nº 186

THE WOODCOCK.

diſtant countries, but only ſhifts from the higher to the lower regions of the atmoſphere ‡. It quits the ſummits of the Andes and Pyrenees on the firſt fall of ſnow, which happens on theſe elevations about the beginning of October; it ſettles below on the gentle acclivities, or advances into the ſunny vales.

The Woodcocks arrive in the night, and ſometimes during the day, in cloudy weather §, always one by one, or two together, but never in flocks. They alight among large hedge-rows, copſes, and tall clumps, and prefer thoſe woods which abound with looſe mould and fallen leaves. They lie concealed the whole day, and lurk ſo cloſe that it requires a dog to put them

‡ "The Woodcock is a bird which reſides in ſummer among high mountains, the Alps, the Pyrenees, Switzerland, and Auvergne, where we have often ſeen it in that ſeaſon. But in autumn it deſcends into the plains and coppices; and ſince Greece abounds with ſuch lofty mountains, we need not wonder that Ariſtotle ſhould not have ſaid that it is migratory. In fact the Woodcock differs from the other birds which entirely leave a country, for it only ſhifts its abode, ſpending the ſummer in the uplands, and the winter in the valleys, where it haunts the perennial ſprings and wet places, while the ſummits of the mountains are frozen, and extracts the worms out of the ground with its long bill; and for this purpoſe it flies in the morning and evening, paſſes the day among the brakes, and comes abroad at night." *Belon.*

§ Willughby.

up,

up, and they often ſpring at the ſportſman's feet. They leave their dark leafy retreats on the approach of evening, and ſpread among the glades, keeping always the little paths. They ſeek the ſoft and wet paſturage by the ſkirts of the wood, and the ſmall meres, where they waſh their bill and feet, which are daubed with earth in ſearching for their food. They have all the ſame œconomy; and we might ſay in general, that the Woodcocks are birds void of character, and that the habits of the individual reſult entirely from thoſe of the ſpecies.

The Woodcock makes a noiſy flapping with its wings when it riſes. In tall groves it ſhoots pretty ſtraight along, but in copſes it is often obliged to deflect its courſe, and dives behind the buſhes to conceal itſelf from the fowler *. Its flight, though rapid, is neither high, nor long ſupported. It ſtops with ſuch promptneſs, as to fall apparently like a dead weight. A few moments after it drops, it runs ſwiftly; but ſoon ſtops, raiſes its head, and caſts a glance all round, before it ventures to lurk in the herbage. Pliny juſtly compares the Woodcock to the Partridge in

* Willughby.

regard

regard to the celerity with which it runs †; for it conceals itſelf in the ſame way, and before the ſportſman reaches the ſpot where he perceives it to alight, he finds that it has already tripped to a great diſtance.

It appears that this bird, though it has large eyes does not ſee well but in twilight, and cannot ſupport a ſtrong light. This ſeems to be evinced by its manner of life, and by its motions, which are never ſo agile as in the dawn, or at the cloſe of the day. And ſo ſtrong is this propenſity to action at the riſe or deſcent of the ſun, that Woodcocks kept in a room have been obſerved to flutter regularly every morning and evening; while during the day or the night, they only tripped on the floor, without attempting to fly. And probably the wild Woodcocks remain ſtill in dark nights; but in moon-light they come abroad in queſt of food. Hence the ſportſmen call the full moon of November, *the Woodcock's moon* (*lune des becaſſes*) ‡, becauſe they are then caught in the greateſt numbers. The ſnares are laid in the evening or the night; and theſe are the *pantenne*, the

† *Ruſticula & perdices currunt.* Plin.

‡ In England, it is uſually termed *the hunter's moon*. T.

ſpring

ſpring or the nooſe: the *pantenne* or *pen-tiere* is a net ſpread between two large trees, in the opening or ſkirt of a wood, where the birds are obſerved to paſs in their evening flight. They may alſo be ſhot in the meres, or on the brooks and fords at ſunſet: the fowler ſits in a cloſe arbour near the ſpot which the Woodcocks frequent; and a little after the cloſe of day, eſpecially when the mild ſouth or ſouth weſt winds blow, they generally arrive, one or two together, and alight on the water, where they may be fired at, with almoſt certain ſucceſs. Yet this ſport is leſs profitable and more precarious than that practiſed by ſprings ſet in their paths. Theſe are ſwitches of hazel, or other flexible and elaſtic wood, driven into the ground, and tied down to a trap that encircles a nooſe of hair or pack-thread: the reſt of the path is incloſed with boughs; or if the ſpring be planted on paſture-ground, brooms or junipers are ſtuck in rows, ſo as to leave only a ſmall paſſage. The bird advances in the track, and being averſe to leap or fly, it bruſhes on the trap, which ſtarting, it is ſeized in the nooſe and lifted into the air by the recoil of the ſwitch. The Woodcock thus ſuſpended ſtruggles much, and the ſportſman muſt make more than

than one progreſs among his ſnares in the evening, and ſtill more towards the end of the night, elſe the fox, a more diligent hunter, informed by the flapping of their wings, will pay a viſit, and carry them off one after another, and, without taking time to eat them, will conceal them in different places, to be devoured at leiſure. The haunts of this bird may be diſcovered by its excrements, which are large, white, and inodorous. To invite it into paſtures where there are no paths, furrows are traced, which it follows in ſearch of the worms that are turned up, and it is entangled in the gins or hair-nooſes placed in the line.

But are theſe not too many ſnares for a bird which can ſhun none? The Woodcock is naturally dull and ſtupid; it is *a very ſottiſh creature* (*moult ſotte béte*) ſays Belon *. It muſt indeed be ſuch, if it can be caught in the way which he relates, and which he calls a *waggery* (*folatrerie*). A man covered with a hood of the colour of dry leaves, walks bent on two ſhort crutches, and approaching gently, he ſtops when the Wood-

* "With us, ſays Willughby, this bird is notorious for its ſtupidity, inſomuch as to have become proverbial." And probably for the ſame reaſon, the Woodcock, as we are told by Dr. Shaw, is called in Barbary, *Hammar el Hadjel*, or the Aſs of the Partridges.

cock

cock ſtops, and reſumes his motion when it does, until he ſees it fixed with its head low; he now ſtrikes his two ſticks againſt each other, and *the bird ſo amuſes* and *befools itſelf*, ſays our old naturaliſt, that the perſon can get ſo near it as to ſlip a nooſe over its neck. Might not this ſtupidity and this facility of diſpoſition induce the ancients to aſſert that the Woodcock had a wonderful attachment to man *? On that ſuppoſition, its affections are very miſplaced, and beſtowed on its greateſt enemy. In fact, it ranges through the woods as far as our farm-hedges and country-houſes: ſo much Ariſtotle remarked †. But Albertus was miſtaken when he ſaid, that it ſeeks the cultivated ſpots and the gardens, to gather ſeeds; ſince neither the Woodcock, nor any bird of that kind, will touch fruit or grain ‡. The ſtraight ſhape of its bill, which is very long and weak at the point, would alone preclude that ſort of food; and in fact, it lives wholly on worms §. It

Ariſtotle, *Hiſt. Anim.* Lib. ix. 26.

† "It is caught by the garden hedges." *Id. Ibid.* "It is alſo ſeen near inhabited places, particularly along hedges." *Olina.*

‡ *In Lib. ix.* Ariſtot.

§ Schwenckfeld. "As ſoon as they enter the woods, they run on the heaps of dry leaves, which they turn over and

It digs in the ſoft ſoil near bogs and ſprings, in looſe paſture-mould and in the wet meadows that ſkirt the woods: it does not ſcrape the earth with its feet; it only turns over the leaves with its bill, and toſſes them briſkly from right to left. It ſeems to diſcover its food by the ſmell ||

and ſcatter to find the worms that lurk underneath. The Woodcocks have this habit in common with the Lapwings and the Plovers, which take the worms by the ſame means under the green herbage. But I have remarked that theſe latter birds, of which I have raiſed ſeveral in my garden, ſtruck the ground with their foot about the holes where the worms were lodged, probably to make them come out by this commotion, and ſeized them often before they were entirely emerged." *Note communicated by M. Baillon, of Montreuil-ſur-mer.*

|| Here is the way in which M. Bowles ſaw Woodcocks feed at St. Ildephonſo, where the Infant Don Louis had a volery filled with all ſorts of birds.

"There was, ſays he, a fountain which flowed perpetually to keep the ground moiſt . . . and in the middle a pine tree and ſhrubs, for the ſame purpoſe. Freſh ſod was brought to them, the richeſt in worms that could be found: in vain did the worms ſeek concealment, when the Woodcock was hungry; it diſcovered them by the ſmell, ſtuck its bill into the ground, but never higher than the noſtrils, drew them out ſingly, and raiſing its bill into the air, it extended upon it the entire length of the worm, and in this way ſwallowed it ſmoothly, without any action of the jaws. This whole operation was performed in an inſtant, and the motion of the Woodcock was ſo equal and imperceptible, that it ſeemed doing nothing. I did not ſee it once miſs its aim; for this reaſon, and becauſe it never plunged its bill up to the orifice of the noſtrils, I concluded that ſmell is what directs it in ſearch of its food." *Natural Hiſtory of Spain,* by G. Bowles, *in 8vo. p.* 454, *&c.*

 rather

rather than by the ſight, which is weak *. But nature has given, at the extremity of its bill, an additional organ, appropriated to its mode of life; the tip is rather fleſh than horn, and appears ſuſceptible of a ſort of touch, calculated for detecting its prey in the mire: and this advantageous ſtructure has been beſtowed alſo on the Snipes, and probably on the Red-ſhanks, the Jadnekas, and other birds which ſearch for food in wet earth †.

The bill of the Woodcock is rough, and almoſt barbed at the ſides near the tip, and hollowed lengthwiſe with deep furrows; the upper mandible alone forms the round point of the bill, projecting over the lower mandible, which is ſomewhat truncated, and fitted below by an oblique joint: it is from the length of its bill, that the name of the Woodcock is derived in moſt languages ‡. The head is equally remarkable; it is rather ſquare than round, and the cranium makes almoſt a right

* *Non illa oculis, quibus eſt obtuſior, & ſi*
Sint nimium grandes, ſed acutis naribus inſtat,
Impreſſo in terram roſtri mucrone . . . Nemeſianus.

† This excellent remark was communicated by M. Hebert.

‡ The Greek name Σκολοπαξ ſignifies ſtake, and ſo does the Hebrew *Kore*: the German *Lang-naſen*, or *Lang-ſchnabel*, (long-noſe, or long-neb) has a ſimilar meaning.

angle

angle at the orbits of the eyes : its plumage, which Ariſtotle compares to that of the *Attagas*, or Red Grous, is too well known to require deſcription ; and the fine effect of the contraſt of light and ſhade, produced by intermingled and broken tints, waſhed with gray, with biſtre, and with umbre, would be tedious and difficult to deſcribe in detail.

We found a gall bladder in the Woodcock, though Belon was perſuaded that it had none ; this bladder diſcharged a liquor by two ducts into the *duodenum* : beſides the two ordinary *cæca*, we perceived a third placed about ſeven inches from the firſt, and which had a diſtinct communication with the inteſtinal canal ; but as we obſerved only one individual, this additional *cæcum* may have been accidental. The gizzard is muſcular, lined with a wrinkled inadheſive membrane ; it often contains ſmall bits of gravel, ſwallowed no doubt along with the earth-worms : the length of the inteſtines is two feet nine inches.

Geſner gives a juſter notion of the bulk of the Woodcock, when he makes it equal to that of a Partridge, than Ariſtotle who compares it to that of a Hen *. That com-

* Ariſtot. *Lib. ix.* 26.

 pariſon

parison however would seem to shew, that the breed of poultry in ancient Greece was much smaller than ours. The Woodcock is always plump; and about the end of autumn it is exceeding fat*. At that season and during the greatest part of the winter, it is reckoned one of the most exquisite dishes †; though its flesh is black and not very tender: that firmness makes it keep long untainted. It is cooked without being emboweled; and its entrails pounded with their contents, make the best sauce for it. It is observed that dogs will not touch this game, and its odour must be offensive to them, for none but spaniels will fetch a Woodcock. The young birds have less *fumet*, but their flesh is whiter and tenderer. They all grow lean as the spring advances, and such as remain during the summer are dry, hard and rank.

* Olina and Longolius say that it is fattened with a paste made of the meal of buck-wheat (*farina d' orzo*), and with figs: in a bird so shy and so fat in its proper season, we cannot imagine how this plan would be either practicable or useful.

† It appears from Olina's account, that, in Italy, the Woodcocks are caught during the whole of the winter: the intense cold which prevails in France at the depth of that season, obliges the Woodcocks to shift their residence a little; yet some of them still remain in the woods, near the warm springs.

It

It is in the end of winter, or in the month of March, that almoſt all the Woodcocks leave our plains, and return to their mountains invited by the pleaſures of love and ſolitude *. They are obſerved to retire in pairs †, and they fly then rapidly, without halting in the night; but in the morning, they conceal themſelves in the woods, where they paſs the day, and in the evening reſume their journey ‡. During the whole of the ſummer, they live in the moſt ſolitary and lofty parts of the mountains, where they breed; as in thoſe of Savoy, of Switzerland, of Dauphiné, of Jura, of Bugey, and of the Voſges. A few remain in the hilly parts of England and of France; ſuch as Burgundy, Champagne, &c. And inſtances may occur of ſome pairs of Woodcocks halting even in our low provinces and neſtling there; detained probably by accidents, and ſurprized by the ſeaſon of love, when at a great diſtance from their proper retreats §. Edwards ſuppoſed, that, like ſo many other birds,

* Belon.

† "They leave England in the beginning of the ſpring, after having paired, and the Cock and Hen fly together." *Willughby.*

‡ Obſervation made by M. Baillon of Montreuil ſur-mer.

§ A Woodcock's neſt was found on the 14th May 1773, on the eſtate of Pont-de-Remy, near Abbeville.

 they

they all advanced into the remoteſt Parts of the north. Probably he was not informed that they removed to the mountains, and that their migrations were of a nature different from the common.

The Woodcock makes its neſt on the ground, like the other birds which do not perch *: in the conſtruction, it employs leaves, or dry herbs, intermixed with ſmall ſticks; the whole artleſsly faſhioned and heaped againſt the trunk of a tree, or under a thick root. It lays four or five eggs, which are oblong, ſomewhat larger than thoſe of a common Pigeon, of a ruſty gray, marbled with deeper and blackiſh waves. One of theſe neſts with the eggs, was brought to us, ſo early as the 15th of April. When the young are hatched, they leave the neſt and run, though ſtill covered with down: they begin even to fly before they have other feathers than thoſe of the wings. They thus make their eſcape, flickering and tripping when diſcovered. The parents ſometimes take a weak one under their throat, and convey it more than a thouſand paces; and the male

* "The Partridges, and other birds which ſeldom fly, neſtle on the ground: of theſe alſo the Sky Lark, the Woodcock, and the Quail, never alight on a tree but on the ground." Ariſtotle, *Lib. ix.* 8.

never

never deſerts the female, until the brood no longer need their aſſiſtance. He is never heard but in his amours, and during the education of the young; for, as well as his mate, he is ſilent the reſt of the year *. During incubation, he ſits conſtantly beſide her; and they ſeem ſtill to ſeek mutual enjoyment by reſting their bill upon one anothers back. Theſe birds, whoſe temper is ſolitary and wild, are diſpoſed to tenderneſs and love: they are even liable to be jealous; for the males ſometimes fight about the female, and peck one another until they fall to the ground. They become not therefore ſtupid and timorous, till after they have loſt the feeling of love, which almoſt ever inſpires courage.

The ſpecies of the Woodcock is univerſally diffuſed: which is remarked by Aldrovandus and Geſner. It is found both in the north and the ſouth, in the Old and the New World: it is known over the whole of Europe, in Italy in Germany, in France, in Poland, in Ruſſia †, in Sile-

* Theſe weak cries are of different tones, paſſing from low to high, *go, go, go, go*; *pidi, pidi, pidi*; *cri, cri, cri, cri*: theſe laſt ſeem to be the expreſſions of anger between ſeveral males aſſembled together. They have alſo a ſort of croaking *couan, couan*, and a certain growling, froũ, froũ, froũ, when they purſue each other.

† Rzaczyncſki.

ſia *, in Sweden †, in Norway ‡, and even in Greenland, where it is called *ſauarſuck*, and the Greenlanders have, according to the genius of their language, a compound name, to expreſs the *Woodcock-fowler* §. In Iceland, the Woodcock conſtitutes a part of the game which there abounds ‖. It occurs, too, in the northern and eaſtern extremities of Aſia where it is common, ſince it has appellations in the languages of the Kamtchadales, the Koriaks, and the Kuriles ** Gmelin ſaw many of them at Mangaſea, on the Jeniſca in Siberia: but theſe were only a ſmall portion of that multitude of Aquatic Birds, which, in the proper ſeaſon, collect on the banks of that river.

The Woodcocks are found alſo in Perſia ††, in Egypt in the neighbourhood of Cairo ‡‡, and theſe are probably what paſs the iſland of Malta in November, con-

* Schwenckfeld.

† *Fauna Suecica.*

‡ Brunnich *Ornithol. Boreal.*

§ *Saurſukſiorpok*, the Greenlandic Dictionary.

‖ Anderſon, *Hiſt. Gen. des Voyages, tom. xviii. p.* 20.

** Among the Kamtſchadales *ſaakouloutch*; among the Koriaks *tobeieia*; and in the Kurile iſlands *petoroi. See* the vocabalaries of theſe languages in *l'Hiſtoire Generale des Voyages*, tom xix, p. 359.

†† Voyage de Chardin, *Amſterdam*, 1711, *tom. ii. p.* 30.

‡‡ Voyage d' Egypte, *par Granger, p.* 237.

veyed

veyed by the north and north-eaſt winds, and which halt not, unleſs detained by contrary winds *. In Barbary, they appear as in the temperate parts of Europe, from October to March †, and it is ſomewhat ſingular, that this ſpecies, which ſeems to be a native of the Frigid Zone, ſhould likewiſe inhabit the Torrid: for Adanſon found a Woodcock on the iſlets of Senegal ‡; and other travellers have ſeen theſe birds in Guinea §, and on the Gold-coaſt ||; Kœmpfer obſerved one at ſea, between China and Japan ¶; and Knox ſeems to have diſcovered them at Ceylon **. Since the Woodcock therefore occupies every climate, we need not wonder that it ſhould occur in America: it is common in the Illinois, and in all the ſouthern part of Canada ††, as well as in Louiſiana; where it is larger than in Europe, which may be aſcribed to the abundance of food ‡‡. It

* Obſervation communicated by the Chevalier Deſhayes.

† Shaw's Travels.

‡ Voyage au Senegal, *p.* 169.

§ Boſman, *Voyage en Guinee*; *Utrecht*, 1705.

|| Hiſtoire Generale des Voyages, *tom. iv. p.* 245.

¶ *Hiſt. Nat. du Japon, tom. i. p.* 44.

** *Hiſt. Gen. des Voyages, tom. viii. p.* 547.

†† *Hiſt. de la Nouv. France*, par le P. Charlevoix, tom. iii. 155.

‡‡ Le Page du Pratz, *Hiſt. de la Louiſiane, tom. ii. p.* 126.

is

is more rare in the northern provinces of America; but the Woodcock of Guiana, known at Cayenne under the name of the *Savanna Woodcock*, ſeems to differ ſo much from ours as to conſtitute a new ſpecies. We ſhall deſcribe it, after enumerating the few varieties of the European ſpecies.

Varieties of the WOODCOCK.

I. THE WHITE WOODCOCK. This variety is rare, at leaſt in our climates * Sometimes its plumage is entirely white; oftener intermingled with ſome waves of gray or cheſnut; its bill is yellowiſh-white; its legs pale yellow, with white nails: that circumſtance would ſeem to ſhew that the whiteneſs is a different ſort of degeneracy from the change of black into white, and much ſimilar to that of the *Blanchards* in Negroes.

II. THE RUFOUS WOODCOCK. The whole plumage conſiſts of different ſhades of rufous, diſpoſed in waves of a darker caſt on a lighter ground. This bird is ſtill more

* One was killed near Grenoble in the month of December 1774: *Letter of M. de Morges, dated Grenoble 29th February 1775.*

rare

rare than the preceding. Both were killed by the king's fowling party in the month of December, 1775, and his majeſty did us the honour of ſending them by the Count d'Angiviller, to be placed in his cabinet of natural hiſtory,

III. Sportſmen pretend to diſtinguiſh two breeds of Woodcocks * *a greater* and *a leſſer:* but as their inſtincts and habits are the ſame, and in every other reſpect they are perfectly ſimilar I ſhall regard the ſmall variation of ſize as only accidental or individual. [A]

* I have frequently remarked that there are two kinds of Woodcocks. The firſt that arrive are the largeſt; their legs are gray, ſlightly inclined to roſe-colour: the others are ſmaller, their plumage ſimilar to that of the Great Woodcock, but their legs are blue. It is remarked that when this little kind is taken in the neighbourhood of Montreuil in Picardy, the Great Woodcock becomes then rare. *Note communicated by M. Baillon of Montreuil-ſur-mer.*

[A] Specific character of the Woodcock, *Scolopax Ruſticola:* "Its bill is ſtraight and tawny at the baſe, it legs cinereous, its thighs feathered, a black ſtripe on its head."

Foreign Bird which is related to the WOODCOCK.

The SAVANNA WOODCOCK.

Scolopax Paludosa. Gmel.

THIS Woodcock of Guiana, though one fourth ſmaller than that of France, has a ſtill longer bill; it is alſo rather taller: its legs and bill are brown; light gray interſected and variegated with bars of black, predominates in its plumage, which is leſs mixed with rufous than in our Woodcock.' With theſe exterior differences, which the climate has perhaps occaſioned, thoſe of its œconomy and habits, which it alſo produces, may be traced, in the Savanna Woodcock: it dwells conſtantly in thoſe immenſe natural meadows where it is never moleſted by men or dogs; it lodges in the bottoms where the mud is collected, and where the herbs are thick and tall; avoiding however thoſe ſhallows where the tide riſes, and makes the water brackiſh. In the rainy ſeaſon, theſe Little Woodcocks

cocks remove to the uplands, and lodge among the grafs: at this time they pair and neftle on the gentle elevations in holes lined with dry herbs. They lay only two eggs; but they have a fecond hatch, and in July, after the rains are over, they again defcend into the bottom grounds, fhifting their refidence from the heights to the vallies the fame way as thofe of Europe. When the Savannas are fet on fire, which is often done in September and October, they efcape in great numbers into the circumjacent tracts, but feem to avoid the woods, and when purfued they never halt, or endeavour to regain their ancient fite; which is contrary to the habits of the European Woodcock. Yet they fpring, like the latter, always under the fportfman's feet; they make the fame effort in rifing, have the fame noify flight, and difcharge their excrement, too, in commencing their motion. If one is fhot, it does not efcape to a diftance, but wheels round until it drops. They are generally flufhed two and two, fometimes three together; and when we fee one we may be fure that the other is not far off. They are heard on the approach of night calling on each other with a raucous voice, much like the low cluck often made by the domeftic hen

ka,

ka, *ka*, *ka*, *ka*. They come abroad at night, and in moon-light they ſit even at the planter's doors. M. de la Borde, who made theſe obſervations at Cayenne, aſſures us, that the Savanna Woodcock is at leaſt as delicate food as the Woodcock of France. [A]

[A] Specific character of the *Scolopax Paludoſa*: "Its bill and legs are brown, its ſtraps and eye-brows black; its body black variegated above with rufous, below partly white."

N.º 187

THE COMMON SNIPE.

The SNIPE.

Le Becaſſine. Buff.

Firſt Species.

Scolopax-Gallinago. Linn. and Gmel. &c.
Scolopax Media. Klein.
Gallinago. Briff.
Gallinago Minor. Aldrov. and Belon

IN the French and German languages the name of the Snipe is the diminutive of the appellation given to the Woodcock; and upon viewing its figure, we ſhould naturally take it for a ſmall ſpecies of that bird: *It would be a little Woodcock*, ſays Belon, *were its habits not different*. It has a very long bill, and a ſquare head; its plumage is alſo ſpeckled the ſame, except that the rufous is leſs intermingled, and that the light gray and the black predominate. But its reſemblance to the Woodcock is confined to externals only, and its natural habits

* In Italian *Pizzardella*: in German *Schnepffe* with the epithets *woffer, heers graſſe (water, lord's, graſs)*: in Swediſh, *Mall Snaeppa, Wald-Snaeppa*: in Daniſh, *Dobbelt Sneppe, Steen Sneppe*: in Norwegian *Trold Ruke*: in Icelandi· *Myr Snippe*: in Poliſh *Bekas, Koſielek, Baranek*: in Turkiſh *Jelve*.

habits are oppoſite. It never frequents the woods; it lives in wet meadows, or among the herbs and oziers which edge the brinks of ſtreams. It ſoars to ſuch height as to eſcape from view, though its cry is ſtill heard; this is a feeble note like the bleat of a goat, *mée*, *mée mée*, which has induced ſome nomenclators to term it the *Flying-goat* *: it alſo vents a ſhort weak whiſtle in taking wing. It differs from the Woodcock as much, therefore, in its diſpoſitions and habits, as it reſembles that bird in its plumage and figure.

In France, the Snipes appear in autumn; they are ſeen ſometimes three or four together, but ofteneſt ſingly. They are fluſhed at a conſiderable diſtance, and fly very nimbly; and after three turnings, they ſhoot onwards two or three hundred paces, or tower aloft out of ſight. The ſportſman can bend their courſe and lead them near him by imitating their voice. They continue the whole winter in our provinces, near the unfrozen fountains and the ſmall contiguous bogs; they retire in great numbers in ſpring, which alſo appears to be the ſeaſon of their arrival in many countries where they breed; as in Ger-

* Klein, Schwenckfeld, Rzaczynſki.

many,

many *, in Silesia † and in Switzerland ‡. But in France, a few Snipes remain the whole summer, and nestle in our marshes: Willughby makes the same remark with regard to England, they build in June, on the ground, beneath some large root of alder or willow, in morasses where the cattle cannot reach: their nest is composed of dry herbs and feathers, and contains four or five oblong eggs, of a whitish colour, with rusty spots. The young ones leave their mansion, as soon as they burst from the shell; they appear ugly and shapeless: the mother nevertheless is affectionate to them, and never deserts them till their long bill is firm enough to enable them to procure easily their food.

The Snipe continually nibbles in the ground, though we can hardly say what it eats: nothing is found in its stomach but an earthy sediment and a watery liquor, which is probably the dissolved substance of the worms on which it feeds; for Aldrovandus remarks that its tongue is terminated as in the Woodpeckers by a sharp point, proper for piercing the worms which it digs out of the mud.

* Aldrovandus.
† Schwenckfeld.
‡ Gesner.

The head of the Snipe has a horizontal balancing, and the tail a motion upwards and downwards; it walks leiſurely, its head erect, without hopping or flickering: but ſeldom is it ſurprized in that ſtate; for it carefully conceals itſelf among the ruſhes and herbage of ſlimy bogs, where the fowlers cannot approach, without the aſſiſtance of a ſort of rackets made of light deal, ſo broad as not to ſink in the mud. And as the Snipe ſprings at a diſtance, with great rapidity, and makes ſeveral inflexions before it ſpins along its courſe; it is one of the moſt difficult birds to ſhoot. It may be caught eaſier with a ſpringe, ſimilar to what is ſet in paths for the Woodcock.

The Snipe has commonly abundance of fat, which is of a delicate flavour; and not apt to cloy like ordinary fat *; it is cooked like the Woodcock without extracting the entrails, and is every where eſteemed delicious game.

Though we ſeldom fail in autumn to find Snipes in our marſhes † the ſpecies is not ſo numerous now as formerly ‡; it is however

* Belon.

† Prodigious numbers of theſe birds are ſeen in the marſhes between Laon, Notre-Dame-de Lieſſe, la Fere, Peronne, Amiens, Calais." *Note communicated by M. Hebert.*

‡ " It is a game ſo frequent in the winter ſeaſon, that we ſee nothing more common in the plains of the inland countries." *Belon.*

ever diffused more universally than that of the Woodcock. It occurs in all parts of the world; and some intelligent voyagers have made this observation *. It has been sent to us from Cayenne, where it is called the *Savanna Snipe* †: Frezier found it in the plains of Chili: it is common in Louisiana ‡, where it resorts near the habitations: and it is equally frequent in Canada §, and at St. Domingo ‖. In the Old Continent, it is found from Sweden ¶ and Siberia ** to Ceylon †† and Japan ‡‡: and we have received it from the Cape of

* "We may remark, that Snipes occur in far more countries of the world than any other bird; they are common in almost the whole of Europe, of Asia, and of America." *Cook's Voyage.*

† Though its flesh is very well tasted, this Guiana Snipe never grows fat, no more than the Woodcock of the same country, and according to M. de la Borde, it lays no more than two eggs. It would seem, that, in all countries where the hatches are repeated, the number of eggs in each is diminished.

‡ Le Page du Pratz, *Hist. de la Louisiane, tom. xi. p.* 127.

§ Nouvelle France, *tom. iii. p.* 155.

‖ M. le Chevalier Lefebre Deshayes remarks, that, a month after their arrival they grow so fat as to appear as unwieldly as the quails: they remain in that island until February.

¶ *Fauna Suecica.*

** Gmelin, *Voyage en Siberie, tom. i. p.* 218, *& tom. ii. p.* 56.

†† Knox, in *l' Hist. Gen. des Voyages, tom viii. p.* 547.

‡‡ Koempfer, *Hist. Nat. du Japon, tom. i. pp.* 112. *&* 113.

Good Hope *. It has ſpread into the remote iſlands in the Southern Ocean †; in the Malouines it was ſeen by Bouganville, who diſcovered that its habits were correſpondent to its undiſturbed ſolitude; its neſt was in the open country, it was eaſily ſhot, it betrayed no ſuſpicion, and made no turnings as it roſe ‡: another proof, that the timid habits of animals which fly before man are impreſſed by fear. The Snipe ſeems even to have averſion to man; for Longolius avers that the Woodcock may be reared and even fatted; but that the experiment never ſucceeded with the Snipe §.

It would ſeem that there is a ſmall kind of Snipes, as of Woodcocks; for beſides the Jack-Snipe, of which we ſhall immediately ſpeak, there are, between theſe and the

* This Snipe of the Cape of Good Hope is ſomewhat larger, with its bill longer, and its legs rather thicker, than ours; yet it is evidently of the ſame ſpecies: it is different from another Snipe, which ſeems indigenous to the Cape, and which we ſhall preſently conſider.

† We found on the northern ſide of Ulietea (an iſland near Otaheite) very deep creeks, which took their riſe from fens filled with vaſt numbers of Ducks and Snipes, ſhyer than we expected: we ſoon learned that the people of the iſland, who are fond of eating them, uſually hunt them. *Forſter.*

‡ *Voyage autour du Monde, par M. de Bougainville, tom. i. p.* 122.

§ In Aldrovandus, *tom. iii. p.* 478.

the common ſort, ſome greater and ſome leſſer. But this difference of bulk, being accompanied with no other, either in the inſtincts or the plumage, ſhews at moſt a diverſity of breed, or only an accidental or individual variety; which however has no connection with ſex, for there is no apparent difference between the male and female in this ſpecies, or in the following*. [A]

* Willughby.

[A] Specific character of the Snipe, *Scolopax-Gallinago*: "Its bill is ſtraight and tuberculated, its legs brown; four brown lines on its front." The Snipes breed conſtantly in the fens of Lincolnſhire, in Wolmar foreſt, and in Bodmyn-downs, and as their neſts are frequently in other parts of Great Britain, it is not improbable that they conſtantly reſide in our iſland. In the breeding ſeaſon, they are obſerved to play over the moors, piping and humming. It is uncertain whether this humming, which is always heard when the bird deſcends, be ventriloquous, or be produced by the motion of the wings.

The JACK-SNIPE.

La Petite Becassine, surnommée La Sourde. Buff *,

Second Species.

Scolopax-Gallinula. Linn. and Gmel.
Gallinago Minima Ray. Will. Klein, Barr.
Gallinago Minor. Briss.
Cinclus. Charleton and Johnston.
The Jack Snipe, Gid, or Judcock. Will †.

THIS Snipe is only half the bulk of the other, and *hence*, says Belon, *the purveyors call it two for one* ‡. It lurks in pools among the reeds, and under the dry bullrushes and flags that have dropped on the brink of the water; and so obstinately does it cling to its concealment, that it will not stir till almost trampled on, and rises from under our feet, as if it heard not the rustling of the approach. For this reason, sportsmen have given it the epithet *deaf*. It flies swifter and more direct than the Common Snipe; its flesh is not inferior, and

* That is, *the Little Deaf Snipe*.

† In Flemish *Hals-Schnepff*: in Danish *Ror-Sneppe*: in Polish *Kisik*.

‡ Our English sportsmen, for the same reason, call it the *Half-snipe*. T.

the

the fat as delicate. But the ſpecies is not ſo numerous, or at leaſt, not ſo widely diffuſed: Willughby remarks, that it is leſs frequent in England: Linnæus omits it in his enumeration of the Swediſh birds; but it is found in Denmark, according to Brunnich. The Jack Snipe has a proportionally ſhorter bill than the Common Snipe; its plumage is the ſame, with ſome copper-reflexions on the back and long ruſty daſhes on the feathers which are laid on the ſides of the back, and which being long, ſilky, and ſomewhat filamentous, have probably given occaſion to the German name *haar-ſchnepffe* or *Hair-ſnipe*.

Theſe birds are ſtationary, and breed in our marſhes; their eggs are of the ſame colour with thoſe of the Common Snipes, but ſmaller, correſponding to their bulk, which exceeds not that of a Lark. They have often been taken for the males of the ordinary kind, and Willughby corrects that popular error, owning at the ſame time, that before comparing them he entertained the ſame notion. Yet Albin has fallen again into that miſtake.

[A] Specific character of the Jack-ſnipe, *Scolopax-Gallinula*: "Its bill is ſtraight and tuberculated, its legs greeniſh, its ſtraps brown, its rump variegated with violet."

The BRUNETTE.

Tringa Alpina. Linn. and Gmel.
{ *Cinclus Torquatus.*
{ *Gallinago Anglicana.* Briss.

WILLUGHBY gives this bird the name of *Dunlin*, probably a diminutive of *dun*. He says that it is peculiar to the North of England. It seems to differ little from the preceding; its belly is blackish waved with white, and the upper surface of the body is spotted with black, and a little white, on a rufous brown ground. In other respects it resembles the Jack-snipe, and must be a contiguous species or perhaps only a variety of it *.

* This bird is really a Sandpiper, and the same with the *Cincle*, which is treated of in the sequel. Brisson led into this mistake, by making two different species of the same bird. T.

Foreign BIRDS which are related to the SNIPES.

The CAPE SNIPE.

First Species.

Scolopax Capensis. Gmel.
Gallinago Capitis Bonæ Spei. Briss.
The Keuvit. Sparrm.

IT is rather larger than the Common Snipe, but its bill is much ſhorter: the colours of its plumage are not quite ſo dark; a bluiſh gray, broken with little black waves, forms the ground of the upper ſurface, which is interſected by a white line that runs from the ſhoulder to the rump; a ſmall black zone marks the top of the breaſt; the belly is white; the head bound with five bars, one ruſty on the crown, two gray on each ſide, then two white, which encloſe the eye and extend behind. [A]

[A] Specific character of the *Scolopax Capenſis*: "Its bill ſtraight, and a line on its top, tawney; a black bar on its breaſt; a white line on each ſide of its back."

The MADAGASCAR SNIPE.

Scolopax Capensis. Var. 3. Gmel.

Second Species.

THE diſpoſition and mixture of the colours on its plumage, make this bird very handſome; the head and neck are rufous, croſſed by a white ſtreak, which paſſes under the eye, with a black ſtreak over it; the lower part of the neck is encircled by a broad black collar; the feathers of the back are blackiſh with feſtoons of gray; ruſty, gray, and blackiſh, are interſected on the coverts of the wings by ſmall undated and cloſe feſtoons; the middle quills of the wing, and thoſe of the tail, are cut tranſverſely by bars varied with that agreeable mixture, and parted by three or four rows of oval ſpots, from a fine light rufous framed in black; the great quills are traverſed with bars alternatively black and rufous; the underſide of the body is white. This Snipe is near ten inches long.

The CHINA SNIPE.

Scolopax Capensis. Var. 2. Gmel.
Scolopax Sinensis. Lath.

IT is rather smaller than the Common Snipe, but taller; its bill is almost as long; its plumage not so dark; it is marbled on the back with pretty large spots and festoons, of dun-gray, bluish, black and light rufous; the breast is decorated with a broad black festoon; the underside of the body is white; the neck is dotted with light gray and rusty; the head is crossed with black and white streaks.

The Madrass Snipe * described by Brisson resembles the above pretty much in its colours, but its hind toe is as long as those before, from which, methinks, the rules of nomenclature should have excluded it from the genus of Snipes.

* *Scolopax Maderaspatana* Gmel.
Gallinago Maderaspatana. Briss.
The Partridge Snipe. Ray.
The Madras nipe. Lath.

Thus dscribed by Brisson: "Above it is blackish, and variegated with fulvous, below white; its throat and the lower part of its neck fulvous, variegated with blackish spots; the upper part of its head marked with a triple longitudinal bar of blackish brown; its back distingu shed by two longitudinal bars of blackish brown: a transverse black bar on the breast; its tail-quills variegated with black, with fulvous, and with gray."

The BARGES.

OF all the volatile beings on which nature has beſtowed ſo much vivacity and grace, and which ſhe ſeem to have diffuſed through the grand ſcene of her productions, to animate and fill up the void ſpace; the Marſh-birds are thoſe to which ſhe has been the moſt ſparing of her favours. Their perceptions are obtuſe, and their mode of ſubſiſtence conſtrains them to ſpend their lives amidſt the fens ſearching in the mud and wet ſoil: and the inhabitants of the primæval ſlime ſeem not to participate in the happy progreſs which all the other productions of the univerſe have made towards perfection, aided and embelliſhed by the induſtry of man.

None of them has the gaiety or the elegance of the rural ſongſters; they cannot toy and frolick, or wanton in harmleſs combats: they only fly from one cold marſh to another; chained to the damp ground, they cannot like the tenants of the grove, ſport amidſt the boughs, or even alight on them: during the day, they keep in the ſhade; and their weak ſight and natural timidity,

timidity, makes them prefer the obſcurity of the night. In ſearching for their food they are alſo indebted to their eyes rather than to their touch and their ſmell Such is the life of the Woodcocks, of the Snipes, and other birds of the marſh; among which the Barges form a ſmall family immediately below that of the Snipes. The ſhape of their body is the ſame; but their bill is ſtill longer, though faſhioned ſimilarly with a blunt ſmooth tip, and ſtraight, or a little bent and ſlightly raiſed. Geſner is miſtaken when he deſcribes their bill as ſharp and proper for darting fiſhes; for they live only on worms, which they extract from the mud. Their gizzard alſo contains little pebbles, moſtly tranſparent like thoſe in the gizzard of the Avoſet. * Their voice is ſomewhat extraordinary, for Belon compares it to the ſmothered bleating of a goat. They are very reſtleſs, ſpring at a great diſtance, and make a ſcream as they riſe. They are rare in countries remote from the ſea, and they delight in ſalt marſhes. They appear regularly on our coaſts, particularly thoſe of Picardy in the

* An obſervation made by M. Baillon on the migratory Godwits of the coaſts of Picardy, and which led him to think that theſe birds and the Avoſet then come from the ſame countries.

month

month of September *: they are ſeen in flocks and heard paſſing at a vaſt height, in the moon-light evenings. Moſt of them halt in the marſhes; and fatigue renders them not ſo fugitive. They ſpring with difficulty, but run like Partridges; and the ſportſman by making a circuit can collect ſo many as to kill ſeveral at one ſhot. They ſtop only a day or two at the ſame place, and it often happens that in the morning not one is to be found in thoſe marſhes where they were ſo numerous the night before. They breed on our coaſts †, and their fleſh is delicate and excellent food ‡.

We divide this genus into eight ſpecies.

* There called the *Taterlas*.

† Obſervation made on the coaſts of Picardy, by M. Baillon of Montreuil-ſur-mer.

‡ Belon.

No 188

THE JADREKA SNIPE.

The COMMON BARGE.

First Species.

Scolopax-Limosa. Linn. and Gmel.
Totanus. Aldrov.
Limosa. Briss.
Fedoa nostra secunda. Ray and Will.
Totanus Cinereus. Barrere.
The Stone Plover. Ray.
The Lesser Godwit. Penn.
The Jadreka Snipe. Lath.

THE plumage is of an uniform gray, except that of the front and throat, where it is rusty-coloured; the belly and rump are white: the great quills of the wings are blackish without, and whitish within; the middle quills and the great coverts, have much white; the two outer feathers are white, and the bill is black at the point, and reddish its whole length, which is four inches; the legs with the naked part of the thighs are four inches and a half; the whole length, from the tip of the bill to the end of the tail, is sixteen inches, and to the extremities of the toes eighteen.

Hebert tells us, that he killed some birds of this kind in Brie. It appears therefore that they sometimes alight in the interior parts

parts of a country, when driven by the violence of the wind. [A]

[A] Specific character of the Jadreka Snipe, *Scolopax-Limosa*: "Its bill somewhat bent back, red at the base; its legs brown, a white spot on its wing-quills, but none on the four first." It inhabits Iceland, Greenland, and Sweden.

The BARKING BARGE.

Second Species.

Scolopax-Totanus. Linn. and Gmel.
Limosa Grisea. Briss.
Totanus. Gesner. Aldrov. and Ray.
Glareola. Klein.
The Barker. Albin.
The Spotted Red-shank. Penn.
The Spotted Snipe. Lath.

Its upper surface is of a brown-gray, fringed with whitish round each feather; those of the tail are striped across with white and blackish. It differs from the preceding also in size, being only fourteen inches long from the point of the bill to the end of the toes.

It inhabits the fens on the maritime coasts of Europe, both those on the Atlantic ocean and on the Mediterranean *. It is found in the salt marshes, and like the

* Albin.

other

other *Barges* it is timid, and flushes at a distance. It seeks its food only during the night *.

* Adanson.

[A] Specific character of the Spotted Snipe, *Scolopax-Totanus*: "Its bill somewhat straight, its legs brown; its eye-brows, its breast, its belly, and its rump white."

The VARIEGATED BARGE.

Third Species.

Scolopax-Glottis. Linn. and Gmel.
{ *Limosa.*
{ *Glottis.* Gesner.
Limosa Grisea Major Briss.
Pluvialis Major. Ray and Will.
The Green legged Horseman. Alb.
The Green-shank. Penn. and Lath.

IF most of the naturalists had not discriminated this from the preceding, we should have regarded it as the same. The colours are alike, the shape entirely similar, and the only difference is that of size, and we have found by experience that in the same species there are sometimes varieties wherein the ill and the legs are often half an inch shorter or longer than usual. This

This *Barge* has its plumage variegated with white, and that colour fringes and encloſes the brown gray of the feathers on the back; the tail is ſtriped with the ſame, and the underſide of the body is white. The Germans give to both the appellation *Meer-haun* (*Sea-hen*); the Swedes call them *glout*. Might not the latter name have led Geſner by a falſe analogy to take theſe birds for the *glottis* of Ariſtotle, which in another place he ſhows to be a Rail? Albin falls into a palpable error, in ſuppoſing this bird to be the female of the Red-ſhank. [A]

[A] Specific character of the Green-Shank, *Scolopax-Glottis*: "Its bill is ſtraight, red at its lower baſe; its body ſnowy below; and its legs greeniſh.

The RUFOUS BARGE.

Fourth Species.

Scolopax Lapponica. Linn. and Gmel.
Limoſa Rufa. Briſſ.
Totanus Fulvus. Barrere.
The Red Godwit. Edw. Penn. and Lath.

IT is nearly as large as the Barker; all the fore-part of the body and the neck is of a fine rufous; the feathers of the upper ſurface of the body are brown and blackiſh,

blackiſh, ſlightly fringed with white and ruſty; the tail is ſtriped tranſverſely with the laſt colour and with brown. This bird is ſeen on our coaſts: it is found alſo in the north, as far as Lapland. It occurs in America, and was ſent from Hudſon's Bay to England. It is another inſtance of thoſe water-fowls which are common to the northern extremity of both continents. [A]

[A] Specific character of the Red God-wit, *Scolopax Lapponica*: "Its bill yellowiſh and ſomewhat bent back, its legs black; the underſide entirely of a ruſty rufous."

The GREAT RUFOUS BARGE.

Fifth Species.

Scolopax Ægocephala. Linn. and Gmel.
Limoſa Rufa Major. Briſſ.
The Godwit, Yarwhelp, or Yarwip. Ray and Will.
The Common Godwit. Penn.

IT is much larger than the preceding; but it has no rufous, except on the neck and the edges of the blackiſh feathers on the back, which are ruſty; the breaſt and the belly are ſtriped acroſs with blackiſh, on a dirty white ground; the length of the bird, from the bill to the

 nails,

nails, is ſeventeen inches. Beſides theſe differences which ſufficiently diſcriminate the two ſpecies, we are informed by an obſerver that they arrive always ſeparately on our coaſts. The Great Rufous *Barge* differs even from the reſt in its œconomy, if what Willughby ſays be true, that it walks with its head erect on the ſandy bare ſhores without ſeeking to conceal itſelf.— It is obviouſly the ſame with the *Barbary Godwit*, deſcribed in Shaw's Travels. [A]

[A] Specific character of the *Scolopax-Ægocephala*: " Its bill is ſtraight, its legs greeniſh, its head and neck tawny; three of its wing-quills black, and white at the baſe.

The RUFOUS BARGE of Hudſon's Bay.

Sixth Species.

Scolopax Fedoa. Linn. and Gmel.
Limoſa Americana Rufa. Briſſ.
The Greater American Godwit or Curlew from Hudſon's Bay. Edw.

THOUGH this bird is more rufous in its plumage than the preceding, and is rather bigger, we cannot help regarding it as a contiguous ſpecies. It is, as Edwards

Edwards remarks, the largest of the genus, being sixteen inches long from the tip of the bill to the end of the tail, and nineteen to that of the toes. All its upper surface is of a rufous brown ground striped across with black; the first great quills of the wing are blackish, the next bay dotted with black; those of the tail are striped transversely with the same colour and with rufous. [A]

[A] Specific character of the *Scolopax-Fedoa*: "Its bill is straight and yellowish, its legs brown, its secondary wing-quills rufous, dotted with black."

The BROWN BARGE.

Seventh Species.

Scolopax Fusca. Linn. and Gmel.
Limosa Fusca. Briss.
The Dusky Snipe. Penn. and Lath.

IT is about the size of the Barker; the ground of its colour is deep brown and blackish, and the feathers of its neck and wings are fringed with small white lines, which give them an agreeable cloudy or scaly appearance; the middle quills of the wing and the coverts are also bordered and

 dotted

dotted with whitiſh on the edges, the firſt great quill ſhows externally only an uniform brown; thoſe of the tail are ſtriped with brown and white. [A]

[A] Specific character of the *Scolopax Fuſca*: "Its bill is bent at the tip; its body black clouded with white; its rump and wings white below: it is twelve inches long, its bill two inches and a quarter. It breeds in the Arctic regions.

The WHITE BARGE.

Eighth Species.

Recurviroſtra Alba. Gmel.
Scolopax Alba. Linn.
Limoſa Candida. Briſſ.
The White Godwit. Edw.
The White Avoſet. Lath.

EDWARDS obſerves that the bill of this bird is bent upwards as in the Avoſet; a character which in ſome degree obtains in all the *Barges* It is nearly as large as the Rufous Barge; its bill black at the tip, and orange the reſt of its length; all the plumage is white, except a tint of yellowiſh on the great quills of the wing and of the tail. Edwards ſuppoſes that the whiteneſs is produced by the cold climate of Hudſon's Bay from which he received it, and that they

they reſume their brown feathers during the ſummer.

It appears that ſeveral ſpecies of *Barges* have ſpread further into America, and have even reached the ſouthern provinces: for Sloane found our third ſpecies in Jamaica; and Fernandez ſeems to indicate two *Barges* in New Spain by the names *Chiquatototl*, and *Elotototl*, the former being like our Woodcock, and the latter lodging under the ſtalks of maize.

[A] Specific character of the *Recurviroſtra Alba*: "It is entirely white, the inferior coverts of its wings duſkiſh, its bill orange, its legs brown."

The HORSEMEN.

Les Chevaliers. Buff.

"THE French, says Belon, seeing a tall little bird, squatting upon its legs as if seated on horseback, denominated it the *Horseman*." It would be difficult to discover another etymology of their name. They are smaller than the *Barges*, but their legs are as long: their bill is shorter, yet fashioned the same. They live in wet marshy places, and also frequent sides of pools and rivers, and even wade to the knees in the water. They run swiftly on the banks and their little body, says Belon, mounted on such tall stilts trips along gayly and nimbly." Their usual food is worms; and in droughts they are contented with the earth insects, and catch beetles, flies, &c.

Their flesh is esteemed *, but it is a very rare dish; for they are no where numerous, and are besides very difficult to approach.

We are acquainted with six species of these birds.

* Belon.

The COMMON HORSEMAN.

First Species.

Tringa Equestris. Lath.
Totanus. Briss.

IT appears to be as large as the Golden Plover, because it is very thickly clothed with feathers; a roperty which belongs to all the Horsemen. It measures near a foot from the bill to the tail, and a little more from the bill to the nails; almost all its plumage is clouded with light gray and rufous; all the feathers are fringed with these two colours, and blackish in the middle; the same white and rusty are finely dotted on the head, spread on the wing, and border its small feathers; the great ones are blackish; the underside of the body and the rump are white. Brisson says that the legs of this bird are pale-red, and he accordingly applies epithets to it, which would better suit the following species, which is perhaps only a variety of the present.

It is from a slight analogy in the colours, that Belon infers this bird to be the *Calidris* of Aristotle. It frequents the sides of rivers

rivers and ſometimes our pools, but more commonly the ſea-ſhore. It is ſeen in ſome of the provinces of France, and particularly in Lorraine: it occurs alſo on all the downs on the Engliſh coaſts; it extends to Sweden *, to Denmark, and even to Norway †.

* *Fauna Suecica.*

† In Daniſh *Rodbeene*: in Norwegian *Lare-Tite, Lare-Titring*. Brunnich.

The RED-SHANK.

Le Chevalier aux Pieds Rouges. Buff.

Second Species.

Scolopax-Calidris. Linn. and Gmel.
Totanus Ruber. Briſſ.
Gambetta. Aldrov.
The Red ſhank, or Pool-ſnipe. Will. Alb. Penn. & Lath.

THE red legs of this beautiful bird the more diſtinguiſh it as more than half of the thigh is naked; its bill is blackiſh at the point, and of the ſame bright red at the root. This Horſeman is of the ſame ſize and figure with the preceding: its plumage is white under the belly, lightly waved with gray and ruſty on the breaſt, and the foreſide of the neck; variegated on the back

No. 189

THE REDSHANK.

back with rufous and blackiſh, by little croſs bars diſtinctly marked on the ſmall quills of the wing; the great quills are blackiſh.

This is certainly the ſpecies which Belon has mentioned by the name *Red-horſeman*, though Briſſon has taken it differently. Ray ſeems to have been no better acquainted with the bird, ſince he conjectures that it is the ſame with the Great Gray Barge.

The Red ſhank is called *Courier* on the Saône. It is known in Lorraine *, and in Orleanois, where however it is rare †, M. Hebert tells us that he ſaw it in Brie in April. It ſits in the ſhallows of pools; it has a diſagreeable voice, and a feeble whiſtle like that of the Snipe. It is known in the territory of Bologna, under the name *Gambetta*, which is diminutive of *gamba*, a leg. It occurs alſo in Sweden ‡, and has probably migrated into the New World. The *Yacatopil* of Mexico, as deſcribed by Fernandez §, appears to

* M. Lottinger.

† Salerne,

‡ *Fauna Suecica*, No. 149.

§ *Yacatopil*, or Stake-bill, is a bird of the bulk of a Wood-pigeon, its bill being four inches long and ſlender, its legs yellow. The colour of its whole body is mixed with white, cinereous, black and brown, it arrives in Mexico, it feeds on worms, muſt be claſſed with the Snipes." Fernandez, *Hiſt. Nov. Hiſp. pp.* 29 and 69.

resemble

resmble it much both in its dimensions and colours. Some species of this kind must even advance further into America, since Dutertre reckons the Horseman among the birds of Guadeloupe, and Labat discovered it among the multitudes that frequent the isle of *Aves*: on the other hand, one of our correspondents * assures us that he saw great numbers of them at Cayenne and Martinico. So that we cannot doubt that these birds are diffused through almost all the temperate and warm countries of the two continents. [A]

* M. de la Borde.

[A] Specific character of the Red-shank, *Scolopax-Calidris:* "Its bill is straight and red, its legs crimson, its secondary wing-quills white."

The STRIPED HORSEMAN.

Le Chevalier Rayé. Buff.

Third Species.

Tringa Striata. Linn. and Gmel.
Totanus Striatus. Briss.
The Striated Sandpiper. Lath.

It is nearly as large as the Common Snipe: all the upper surface is striped on a gray ground intermixed with rusty, with

with blackiſh ſtreaks diſpoſed tranſverſely; the tail is interſected with the ſame on a white ground; the neck has the ſame colours, except that the brown daſhes are laid along the ſhaft of the feathers: the bill is black at the tip; but, at the root, it is of a pale red, as well as the legs. To this ſpecies we ſhall refer the *Spotted Horſeman* of Briſſon †, which appears to be only a minute variety. [A]

† Above, it is clothed with feathers blackiſh in the middle, and ruſty-gray at the edges; below, white, variegated with blackiſh ſpots; its rump and its lower belly, bright white; its lateral tail-quills ſtriped tranſverſely with white and blackiſh; its legs red." *Totanus Nævius*. Briſſon.

[A] Specific character of the *Tringa Striata*: "The baſe of its bill and its legs are bright yellow; its tail-quills white, ſtriped with brown; many of its wing-quills white."

The VARIEGATED HORSEMAN.

Fourth Species.

Tringa Ochropus. Var. Gmel.
Tringa Littorea. Linn.
Totanus Cinereus. Briſſ.
Calidris Nigra. Belon. Aldrov. Johnſt. and Charl.
Charadrius Nigricans. Barrere.
The Shore Sandpiper. Pen. and Lath.

Its colours conſiſt of blackiſh, rufous, and gray; the blackiſh covers the upper ſide of the head and the back, and the edges of theſe feathers are marked with the rufous; the wings, too, are blackiſh, and fringed with white or ruſty; theſe tints are intermingled with gray on all the forepart of the body; the legs and the bill are black, it is as large as the Red-ſhank, but its legs are not ſo tall.

It appears that this bird builds its neſt very early in the ſeaſon, and that it returns into our provinces before the ſpring: for Belon ſays, that in the end of April the young ones were brought to him, and that their plumage then reſembled much that of the Rail, and that " he had not been accuſtomed to ſee theſe Horſemen, but in winter." They do not breed equally on on all the coaſts of France; for example, we are aſſured, that they only pay a tranſient viſit in Picardy: they are driven thither in the month of March by the north eaſt wind along with the *Barges*: they make but a ſhort ſtay, and do not repaſs till the month of September. They have ſome habits ſimilar to thoſe of the Snipes, though they come leſs abroad during the night, and appear oftener in the day-time. They are caught alſo by the ſpringe *. Linnæus ſays that this ſpecies is found in Sweden. Albin, from an inconceiveable miſtake, calls it, *a White Heron*, though the greateſt part of its plumage is black, and though it has not the ſmalleſt reſemblance to the Heron.

* M. Baillon, who communicated to us theſe facts, adds the following obſervation on one of theſe birds which he kept. "I preſerved one of theſe birds laſt year in my garden more than four months. I remarked that in time of drought, it caught flies, beetles, and other inſects, no doubt for want of worms. It alſo ate bread ſoaked in water, but the maceration needed to be continued a whole day. Moulting gave it, in the month of Auguſt, new feathers to the wings, and it eloped in September. It was grown very familiar, inſomuch that it followed the gardener, when he brought it food. If it ſaw him break off the leaf of a plant, it ran to pick up the worms that were diſlodged: as ſoon as it had eaten, it repaired to waſh itſelf in a bowl filled with water. I never ſaw it with dry earth on its bill or its legs. This attention to cleanlineſs is common to all the vermivorous birds."

The WHITE HORSEMAN.

Fifth Species.

Scolopax Candida. Gmel.
Totanus Candidus. Briff.
The White Red ſhank, or Pool-ſnipe. Edw. and Lath.

THIS bird occurs at Hudſon's Bay. It is nearly the bulk of the firſt ſpecies; all its plumage is white, and its bill and legs are orange.

Edwards thinks that it is one of thoſe birds whoſe plumage turns white from the influence of an arctic winter, and that the brown colour returns in ſummer; and in the figure, which that author gives, a tint of that colour appears on the great quills of the wings and of the tail, and marks the upper ſurface with ſmall waves.

The GREEN HORSEMAN.

Sixth Species.

Rallus Bengalensis. Gmel.
Totanus Bengalensis. Briss.
The Bengal Water-Rail. Alb.

ALBIN describes this bird; but the figure which he gives is a very bad one: we may trace, however, the bill and legs of the Horseman. It has a green tinge on the back and the wing, except the three first quills, which are purple and intersected by orange spots: there is some brown on the neck and the sides of the head, and white on the crown, as well as on the breast.

The RUFF AND REEVE*

Les Combattans, vulgairement Paon de Mer. Buff.
Tringa Pugnax. Linn. and Gmel.
Pugnax. Briss. and Brun.
Glareola Pugnax. Klein.
Philomachus. Mœhr.

THESE birds are well entitled to the appellation of *combattants*, for they not only contend with each other in single rencounters, but they advance to battle in marshalled ranks †: and these hostile armies are composed entirely of the males, which, in this species, are more numerous, it is said, than the females: the latter wait the issue of the conflict, and become the prize of the victors. Love is then the source of these contentions, which nature seems to countenance by the disproportion between the number of the males and that of the females ‡.

Every spring these birds arrive in great flocks on the coasts of Holland; Flanders and

* In German *Kampfshoehulein*: in Flemish *Kemperkens*: in Swedish and Danish *Brushane*: and in Polish *Ptak-Bitny*.
† Klein.
‡ Aldrovandus.

N.° 190

THE RUFF, IN THE SEASON OF LOVE.

and England; and in all these countries they are believed to come from the north. They are known also on the shores of the German-ocean, and are numerous in Sweden, particularly in Scania *: they occur likewise in Denmark and in Norway †; and Muller says that he thrice received them from Finmark. It is uncertain where they spend the winter ‡. Since they appear regularly on our coasts in the spring, and stay two or three months, it would seem, that they seek the temperate climates; and if observers had not assured us, that they come from the north, we might justly draw the opposite inference, that they arrive from the south. I should therefore presume that it is the case with these birds as with the Woodcocks, which are said to come from the east, and return to the west or the south, but which only descend from the mountains to the plains and again retire to their heights. It is even probable that the Ruffs remain in the same country, only shifting to different parts of it according to the change of seasons; and if their battles be seen only in the spring, it is

* *Fauna Suecica.*

† *Zoolog. Danic.* p. 24.

‡ Charleton says, that they annually arrive in the fens of Lincoln, and, after three months, retire he knows not whither.

 probable,

probable, that at other times, they paſs unobſerved, intermingled perhaps with the Duſky Sandpipers, or the Horſemen, to which they bear great analogy.

The Ruffs and Reeves are nearly the ſtature of the Red-ſhanks, though their legs are not quite ſo tall; the bill has the ſame form, but ſhorter. The females are commonly ſmaller than the males *, and reſemble them in their plumage, which is white with a mixture of brown on the upper ſurface. But the males in the ſpring are ſo different from one another, that each might be taken for a bird of a diſtinct ſpecies: of more than a hundred, compared by Klein at the houſe of the governor of Scania, there were only two preciſely ſimilar; they varied either in ſhape or in the bulk of their ſwelled ruff round the neck. The feathers which compoſe that bunch appear in the beginning of the ſpring, and remain only during the ſeaſon of love. But beſides the exuberance of growth at that time, the ſuperabundance of organic molecules diſplays itſelf alſo in the eruption of fleſhy turgid pimples on the foreſide of the head and round the eyes †. This double pro-

* Rzaczynſki.

† Linnæus, *Fauna Suecica*.

duction

duction implies great prolific powers in the Ruffs. "I know of no bird, M. Baillon writes us, in which the appetite of love is more ardent; none whose testicles are so large in proportion, each being near six lines in diameter, and an inch or more in length; the other organs of generation are equally dilated in the season of its amours. We may judge of the impetuosity of their hostile assaults, from the strength of that passion which blows up the fires of jealousy and rivalship. I have often followed these birds in our marshes (in Lower Picardy) where they arrive in the month of April with the Horsemen, but in smaller numbers. Their first object is to pair, or rather to contend for the females, whose feeble screams rouse and exasperate the antagonists; and the battle is long, obstinate, and sometimes bloody. The vanquished betakes himself to flight, but the cry of the first female which he hears, dispels his fears and awakens his courage; and he again renews the conflict, if another opponent appears. These skirmishes are repeated every morning or evening until the departure of the birds, which happens sometime in May; for only a few stragglers remain, and their nests are never found in our marshes."

This accurate and very intelligent observer remarks, that they leave Picardy with a ſouth, or ſouth or ſouth-eaſt wind, which carries them to the Engliſh coaſts, where they breed in very great numbers, particularly in the fens of Lincolnſhire. In that county, they afford conſiderable ſport; the fowler watches the inſtant when they are fighting, and throws his net over them*. They are fattened for the table with milk and crumbs of bread; but, to keep them quiet and peaceful, they muſt be ſhut up in a dark room, for whenever the light is admittted they fall a quarrelling †. Nor can confinement ever eradicate the ſeeds of diſcord; and in their voleries, they bid defiance to all other birds ‡, and if there were only a bit of green turf, they will fight for the poſſeſſion of it §; as if they gloried in their combats, they ſeem moſt animated in the preſence of ſpectators ‖. The tuft in the males is not a warlike

* Willughby.

† *Id.*

‡ The Chineſe have many birds which they call *fighters*, which they rear not for the ſake of ſong, but to exhibit their rancorous battles. *See* Hiſt. Gen. des Voy. *tom. vi. p.* 487. However theſe are manifeſtly not our Ruffs, ſince they are not larger than Linnets.

§ Klein.

‖ Willughby.

ornament,

N.o 191

THE RUFF IN MOULTING.

ornament only, it is a ſort of defenſive armour which wards off the blows: its feathers are long, ſtiff and cloſe. They briſtle in a threatning manner when the bird makes an attack, and their colours form the chief difference between the individuals. In ſome, theſe feathers are rufous, and in others, gray; in ſome, white, and in others, of a fine violet black, broken with rufous ſpots: the white complexion is moſt rare. In its form, too, this tuft is as variable as in its colours, during the whole time of its growth *.

This beautiful ornament drops in a moult which theſe bird undergo about the end of June; as if nature reſerved her decorations and armour for the ſeaſon of love and of war: the vermilion tubercles grow pale, and obliterate, and their place becomes occupied by feathers: the males are no longer to be diſtinguiſhed from the females, and they all abandon the places

* Of the eight figures which Aldrovandus gives from drawings ſent him by the Count d'Aremberg from Flanders, the one appears to be the female, five others males of moulting or of the growth of their ruff; and the eighth, which Aldrovandus himſelf thought to have ſomething monſtrous, or at leaſt foreign to the ſpecies of the Ruff, is nothing but a bad figure of the Horned Grebe, which this naturaliſt was ignorant of; we ſhall treat of it in the ſequel.

where they bred and hatched. They build in companies like the Herons; and that property alone induced Aldrovandus to claſs theſe birds together; but the ſtature and conformation of the Combatants remove them far from all the ſpecies of Herons. [A]

[A] Specific character of the Ruff and Reeve, *Tringa Pugnax*: "Its bill and legs are rufous, its three lateral tail-quills ſpotleſs, its face marked with fleſhy granulated *papillæ*." Theſe birds appear about the end of April in Lincolnſhire, in the iſle of Ely, and in the Eaſt Riding of Yorkſhire. They are caught with a net about forty yards long, and about ſeven or eight feet high, propped in an inclined poſition near the reeds, where the fowler lurks until the birds, enticed by a ſtale, alight under the net; he then pulls a cord, which lets it fall and ſecures them. They are fattened with milk, hempſeed, and ſometimes with boiled wheat; and to haſten the proceſs, ſugar is frequently added. They are then ſold for half-a-crown a-piece; and it requires judgment to diſcern when they have attained the utmoſt pitch of fatneſs: for if the regimen be longer continued, they will ſicken and pine away. At this critical period they ſell at half-a-crown a-piece. The method of killing them is to ſever the head with ſciſſars; they ſtream a profuſion of blood" They are dreſſed like the Woodcocks, and eſteemed moſt delicious.

The MAUBECHES.

THESE birds may be ranged after the Horſemen and before the Green Sand-piper: they are rather larger than the latter, but ſmaller than the former: their bill is ſhorter, their legs not ſo tall, and their ſhape is rounder than that of the Horſe-men: their habits muſt be the ſame, thoſe at leaſt which depend on their ſtructure and their haunts; for they equally frequent the beach. We know nothing more of their œconomy, though we can enumerate four different ſpecies.

The COMMON MAUBECHE.

Firſt Species.

Tringa-Calidris. Linn. and Gmel.
Calidris. Briſſ.
The Duſky Sand-piper. Lath.

IT is ten inches from the point of the bill to the nails, and a little more than nine inches to the end of the tail; the feathers on the back, on the upper ſide of the

the head, and on the neck, are blackish brown, edged with light chesnut: all the forepart of the head, of the neck, and of the back, is light chesnut; the nine first quills of the tail are deep brown above on the outside; the four next the body are brown, and the intermediate ones are brown gray, and edged with a narrow hem of white. The *Maubeches* in general have the under part of the thigh naked, and the mid-toe connected as far as the first articulation to the outer toe, by a portion of membrane. We cannot, with Brisson, refer to this bird the *Rusticula Sylvatica* of Gesner, which "is larger than the Woodcock, and equal to a common Hen." It would be difficult to class that with any known species: and we may save ourselves the trouble, since Gesner acquaints us that he lays little stress on those of his descriptions which were made from rude and apparently inaccurate drawings. [A]

[A] Specific character of the Dusky Sandpiper, *Tringa-Calidris*: "Its bill and legs are blackish, its body olive below, its rump variegated."

The SPOTTED MAUBECHE.

Second Species.

Tringa Nævia. Gmel.
Calidris Nævia. Briſſ.
Glareola Caſtanea. Klein.
The Freckled Sandpiper. Penn. and Lath.

THIS is diſtinguiſhed from the preceding, becauſe the brown aſh colour of its back and ſhoulders is variegated with conſiderable ſpots, ſome rufous, others blackiſh bordering on violet: it is alſo ſomewhat ſmaller.

The GRAY MAUBECHE.

Third Species.

Tringa Griſea. Gmel.
Calidris Griſea. Briſſ.
The Griſled Sandpiper. Lath.

THIS bird is rather larger than the preceding, but ſmaller than the firſt ſpecies. The ground of its colour is gray; the back is entirely of that colour; the back is gray waved with whitiſh; the feathers

thers of the upper ſide of the wings, and thoſe of the rump, are gray and bordered with white; the firſt of the great quills of the wing are blackiſh brown, and the foreſide of the body is white, with ſmall black zig-zag ſtreaks on the flanks, the breaſt, and on the forepart of the neck.

The SANDERLING.

Fourth Species.

Charadrius Calidris. Linn. and Gmel.
Tringa Arenaria. Ray and Will.
Calidris Griſea Minor. Briſſ.
The Sanderling, or Curwillet. Alb. Will. Penn. and Lath.

THIS is the ſmalleſt of the genus, not exceeding ſeven inches in length. Its plumage is nearly the ſame with that of the preceding, except that all the foreſide of its neck, and the underſide of its body, are very white. Theſe birds fly in flocks, and light on the beach. Willughby gives them four toes to each foot; but Ray, who ſeems however to copy that naturaliſt's deſcription, aſſigns only three, which would rather characterize the Plover. [A]

[A] Specific character of the Sanderling, *Charadrius Calidris*: "Its bill and legs are black, its ſtraps and its rump are grayiſh, its body ſpotleſs, white below." It is common on the Corniſh coaſts, where it is called *Curwillet*.

The GREEN SANDPIPER.

Le Becasseau. Buff.
Tringa-Ochropus. Linn. and Gmel.
Tringa. Aldrov. Gesner. &c.
Cinclus. Belon.
Glareola. Klein *

† It is as large as a common Snipe, but not so long shaped; its back is rusty ash colour with small whitish drops on the edges of the feathers; the head and neck are of a softer cinereous, mixing in streaks with the white of the breast, which extends from the throat to the stomach and the belly; the rump is of the same white; the quills of the wing are blackish, and agreeably spotted with white below ‡; those of the tail are striped across with blackish and with white; the head is square, like that of the Snipe, and the bill is of the same form on a diminished scale.

The Green Sandpiper frequents the sides of water, and particularly running brooks: it runs among the gravel, or skims along

* In Italian *Gambettola, Giarola,* and *Pivinello.* Also *Cul-bianco*; and hence the French *Cul-blanc* or white arse.

† Some strictures on the nomenclators are here omitted T.

‡ Belon.

the

the surface. It utters a screams as it rises and flies, beating its wings with distinct and separate strokes. When pursued it sometimes dives into the water. The Ring-tails often chase it, and take it by surprize while it is reposing by the edge of the stream, or is occupied in the search of its food: for the Green Sandpiper has not the security of birds that go in flocks, which commonly appoint one of their number to watch, as a centinel, the common safety. It lives solitary in a small district, which it selects by the banks of a river, or on the sea-shore *; and there it remains constantly, without roving, to any considerable distance. But this lonely savage mode of life does not extinguish its sensibility; at least its voice has a manifest expression of sentiment: it is a gentle sweet whistle, modulated, with accents of languor, which being diffused over the placid surface of the water, or mingled with its murmur, inspires reflection and tender melancholy. It would appear, that this is the same bird with what is called the *Sifflasson* (*the Whistler*) on the Lake of Geneva, where it is caught by the call, with limed bull-rushes. It is known also on the

* Willughby.

Lake of Nantua, where it is terminated the *Pivette*, or Green foot. It is ſeen likewiſe in the month of June on the Rhone and the Saone; and in autumn among the gravel on the Ouche in Burgundy. Theſe birds occur, too, on the Seine; and it is remarked that though ſolitary the whole ſummer, they form ſmall bodies of five or ſix in the time of their paſſage, and are heard in the air in ſtill nights. In Lorraine, they arrive in the month of April, and retire in July *.

Thus the Green Sandpiper, though attached to the ſame place during the time of its ſtay, paſſes from one country into another, even in thoſe ſeaſons when moſt other birds are detained by the duties of incubation. Though ſeen two thirds of the year in the coaſts of Lower Picardy, it cannot be ſaid to breed there: it is called the *Little Horſeman* in thoſe parts †; it haunts the mouths of rivers, and follows the tides; it picks up the little fry of fiſh, and worms, on the ſand which is left bare and covered alternately by a thin ſheet of water. The fleſh of the Green Sandpiper is very delicate, and even ſuperior in flavour to that

* Obſervations of M. Lottinger.

† Obſervations on the birds of our Weſt coaſts, communicated by M. Baillon.

of

of the Snipe, according to Belon, though it has a ſlight odour of muſk. As it perpetually wags its tail in walking, naturaliſts have applied to it the name *Cinclus*, whoſe primitive ſignifies to ſhake or agitate*: but that character belongs as much to the common Sandpiper, or to the Purre; and a paſſage of Ariſtotle proves clearly, that the term does not correſpond to the Green Sandpiper. That philoſopher denominates the three ſmalleſt Marſh Birds τριγγας σχοινικλος, κιγκλος: "Of theſe, ſays he, the *Cinclus* and *Schœniclus* are the ſmalleſt, and the *Trinjas* the largeſt, being equal to a Thruſh."† The *Tringas* is therefore indicated by its bulk to be the ſame with the Green Sandpiper; but we have not *data* to decide whether the two others correſpond to the common Sandpiper, or to the Purre, or to our little *Cincle*. But nothing can equal the confuſion in which nomenclators have involved this ſubject: ſome term the Green Sandpiper the *Water-hen*, others the *Sea-partridge*; ſome, as we have ſeen, call it by the name *Cincle*, but the greater number apply the term *Tringa*; perverting its ſignification however by making it generic.

* Κιγκλος, from κιγκλιζω, and this from κινεω, to move.
† Hiſt. Animal. *Lib. viii.* 4.

Hence

Hence the profusion of epithets and phrases that have been employed, and the multitude of inaccurate figures and of vague references: and Klein justly laments the impossibility of reconciling the chaos of descriptions which abound in the works of authors, who have blindly copied and compiled, without consulting nature, [A]

[A] Specific character of the Green Sandpiper, *Tringa Ochropus*: "The tip of its bill is pointed, its legs greenish, its back brown-green; its belly and its utmost tail-quills, white."

The COMMON SANDPIPER.

La Guignette. Buff.
Tringa Hypoleucos. Linn. and Gmel.
Guinetta. Briff.
Tringa Minor. Ray.
Gallinula Hypoleucos. Johnft.

It exactly refembles the preceding in its form and plumage, only fmaller. Its throat and belly are white; its breaft fpotted with gray dafhes on white; its back and rump gray, not fpotted with white, but flightly waved with black, with a fmall ftreak of that colour on the fhaft of each feather; and upon the whole, there is a reddifh reflection; the tail is a little longer and more fpread, than that of the Green Sandpiper: it likewife wags its tail as it walks; and hence fome naturalifts have termed it *Motacilla*, which has already been beftowed on a multititude of fmall birds.

The Sandpiper lives folitary by the verge of water, and haunts the fandy ftrands and fhores. Many of thefe birds are feen near the fources of the Mofelle in the Vofges, where they are called *Lambiche.* It leaves that country in the month of July, after having reared its young.

* In German *Fyfterlin:* in Swedifh, *Snaeppa.*

It

N°. 192

THE SANDPIPER.

It ſprings at a diſtance, and utters ſome ſcreams *; and it is heard during the night to cry on the beach with a wailing voice: and the ſame property ſeems to belong to the Green Sandpiper †.

Both theſe ſpecies advance far into the north ‡, and have thence migrated into the cold and temperate parts of the New Continent: in fact, a Green Sandpiper which was ſent to us from Louiſiana, ſeemed to differ little or nothing from that of Europe [A]

* Willughby.

† The *Pilvenckegen*, according to Willughby, repreſented by Geſner as a moaning bird.

‡ *Fauna Suecica*, Nos, 147, and 152.

[A] Specific character of the Sandpiper, *Tringa Hypoleucos*: "Its bill is ſmooth, its legs livid; its body cinereous, with black daſhes, below white."

The SEA PARTRIDGE.

THE name *Partridge* has very improperly been applied to this bird; since the only relation consists in a small resemblance of the bill, which is pretty short, convex above, compressed at the sides, and curved near the point, as in the gallinaceous tribes. In the form of its body and the fashion of its feathers, it bears more analogy to the Swallows; for its tail is forked, its wings have great extent and stretch to a point. Some authors have called it *Glareola*, alluding to its frequenting the strands by the sea-shore. It feeds chiefly on worms and Aquatic insects. It also haunts the brinks of brooks and rivers, as on the Rhine near Strasburg, where according to Gesner it has the German appellation *Kappriegerle*. Kramer denominates it *Pratincola* (*Prati-incola*) because he saw many of them in the extensive meadows which border on a certain lake in Lower Austria. But whether it inhabits the verge of rivers and lakes, or the sea-shore, it universally prefers the strands and sandy channels, to the muddy bottoms.

We

We know four ſpecies or varieties of theſe Sea-partridges which ſeem to form a ſmall diſtinct family amidſt the numerous tribe of little Shore-birds.

The GRAY SEA-PARTRIDGE.

Firſt Species.

Hirundo-Pratincola. Linn.
Glareola-Auſtria. Gmel.
Pratincola. Klein.
Glareola. Briſſ.
Hirundo Marina. Ray. Will. and Johnſt. &c.
The Auſtrian Pratincole. Lath.

THIS, with the following ſpecies, appears ſometimes, though ſeldom, on the rivers in ſome of the provinces of France, particularly in Lorraine, where Lottinger aſſures us he has ſeen it. All the plumage is gray, tinged with rufous on the flanks and on the ſmall quills of the wings; the throat only is white incloſed by a black rim; the rump is white, and the legs red: it is nearly as large as a Blackbird. The Sea-ſwallow of Aldrovandus, which, in other reſpects, is much analogous to this ſpecies, appears to form a variety, it having, according to that naturaliſt, very black legs.

The BROWN SEA-PARTRIDGE.

Second Species.

Glareola Senegalenſis. Gmel. and Briſſ.
Tringa Fuſca. Linn.
The Senegal Pratincole. Lath.

THIS ſpecies, which is found in Senegal and is of the ſame ſize with the preceding, differs however in being entirely brown. We are inclined to think that this variation reſults from the influence of climate.

The GIAROLE.

Third Species.

Glareola Nævia. Gmel. and Briſſ.
Gallinula Melampus. Aldrov. Geſn. Ray and Klein.
The Spotted Pratincole. Lath.

GIAROLA is the name which this bird receives in Italy. Aldrovandus properly refers to it the *Melampus* (*Black foot*) of Geſner, all thoſe of this genus being diſtinguiſhed by their black legs. The German appellation *Rotknillis* (*red-clouded*) alludes

alludes to the reddiſh ground of its plumage, ſpotted with whitiſh or brown; the wing is cinereous, and its quills black.

The COLLARED SEA-PARTRIDGE.

Fourth Species.

Glareola Auſtriaca. Var. 1. Gmel.
Glareola Torquata Briſſ.

THE German name of this bird *Riegerle*, implies that it is perpetually in motion. In fact, when it hears any noiſe, it is alarmed, and runs or flies away with a feeble ſhrill cry. It frequents the ſhores, and its habits are perhaps the ſame with thoſe of the Common Sandpiper. But if we admit the accuracy of Geſner's figure, both the form of its bill and the colours of its plumage, import its relation to the Sea-partridges: the back is cinereous, and alſo the upper ſurface of the wing, of which the great quills are blackiſh; the head is black, with two white lines on the eyes; the neck is white, and a brown circle ſurrounds it below like a collar; the bill is black, and the legs yellowiſh. It is one of the ſmalleſt ſpecies: Schwenckfeld ſays that it breeds on the ſandy brinks of rivers, and lays ſeven oblong eggs; he adds that it runs very faſt, and, in the ſummer nights, utters a little ſcream *tul, tul,* with a ringing voice.

The SEA-LARK.

Tringa Cinclus. Linn. and Gmel.
The Least Snipe Sloane and Brown.
The Purre, or Stint *. Penn. and Lath.

THOUGH this bird has the name of Lark, it bears ſcarce any reſemblance to it, except in its bulk and in the plumage of its back †: in its form and its habits, it is entirely different. It lives by the verge of water; and never leaves the ſhores; the under part of its thigh is naked, and its bill ſlender, cylindrical, and obtuſe, as in the other *Scolopacious* birds, and this is only ſhorter in proportion than that of the Jack-ſnipe, which it reſembles in its air and figure.

It prefers the ſea coaſt, though it alſo frequents the rivers; it flies in flocks often ſo thick that a number may be killed by a ſingle ſhot; and Belon expreſſes his wonder to find the markets on our coaſts ſo well ſupplied with theſe birds. According to him, the fleſh is better than that of a common Lark; but if kept for any length

* In Suſſex, it is called the Ox-eye, according to Ray. T.
† Belon.

of

of time, it contracts an oily taste. When one is killed, the rest crowd round the sportsman, as if to protect their companion. Faithful to each other, they give a mutual scream in rising, and in a body skim along the surface of the water. At night, they are heard to call and cry on the strands and little islets.

They assemble in autumn; those which had separated to breed re-unite with their new families which usually consist of four or five young. The eggs are very large in proportion to the size of the bird: they are dropped on the naked sand. The Common and Green Sandpipers have the same habits, and build no nest. The Sea Lark procures its prey along the shore, walking, and perpetually wagging its tail.

These birds pass into other countries; and it appears that on some of our coasts, they are migratory only. Of this we are assured, at least with regard to Lower Picardy, by an excellent observer *: there they arrive in the month of September with the westerly winds, and make only a transient halt; they will not suffer a person to approach to them nearer than twenty paces, which would make us conjecture that they are not

* M. Baillon.

hunted

hunted in the country from whence they come.

They muſt alſo have penetrated far into the north, and have paſſed into the other continent; for the ſpecies is found to be ſettled in Louiſiana * in the Antilles †. Jamaica ‡, St. Domingo and in Cayenne §. The two Sea-larks from St. Domingo which Briſſon has given ſeparately, appear to be only varieties of the European ſpecies. In the Old Continent they are ſpread from north to ſouth; for according to Kolben it occurs in the Cape of Good Hope, and Willughby and Sibald repreſent the *Stint* as a native of Scotland. [A]

* Le Page du Pratz; *Hiſt. de la Louiſiane, tom. ii. p.* 118.

† The Sea-larks and other ſmall Sea-birds are aſtoniſhingly numerous in all the ſalt marſhes, *Dutertre*, tom. xi. p. 277.

‡ Sloane and Browne.

§ "Every year theſe birds are ſeen in Cayenne, and on the whole coaſt; they aſſemble at ſpring tides, and ſometimes in ſuch numbers, that the banks of the rivers where the influx reaches are covered with them, either on the ground or in the air; their flocks are ſo thick, that a perſon will ſometimes kill forty or fifty of them at one ſhot. The inhabitants of Cayenne alſo hunt them at night on the ſands, where theſe birds eat the little worms left by the ſea. They ſometimes perch on the mangroves by the water's edge; their fleſh is very good eating. In the rainy ſeaſon, they are very numerous in St. Domingo and Martinico; but it is uncertain how they breed, or where they lay their eggs." *Remarks made by M. de la Borde, King's phyſician at Cayenne.*

[A] Specific character of the Purre or Stint, *Tringa-Cinclus*: "Its bill and legs are black, its ſtraps white, its tail and rump gray and brown."

The CINCLE.

Tringa Alpina. Linn. and Gmel.
{ *Cinclus Torquatus.*
{ *Gallinago Anglicana.* Briss.
The Dunlin. Penn. and Lath.

ARISTOTLE has applied the name *Cinclos* to the least of all the Shore-birds; and we have adopted it. It appears to be subordinate to the preceding species; rather smaller, not so tall; it has the same colours, with this single difference, that they are more distinct; the dashes on the upper surface are more finely traced, and there is a zone of spots of the same colour on the breast: which has induced Brisson to call it the *Collared Sea Lark*. In other respects, it has the habits of that bird, and it often associates with it. Its tail has a sort of tremulous motion, which Aristotle seems to ascribe to his *Cinclos*. But we have not been able to discover in it the other properties which he mentions; viz. that being once caught it grows very tame, though it shews much cunning in avoiding snares. With regard to the long and obscure dissertation

tation of Aldrovandus on this ſubject, we can derive nothing more than that the Italians apply the names *Giarolo* and *Giaroncello* to the Dunlin and the Stint. [A]

[A] Specific character of the *Tringa Alpina*: "It is brown brick-coloured, its breaſt blackiſh, its tail-quills whitiſh cinereous, its legs duſkiſh."

To the hiſtory of the Land-birds we ſhall ſubjoin the following particulars, tending to throw light on the ſubject.

The huſbandmen of antiquity were directed in their agricultural labours by the riſing of ſome noted conſtellations, by the appearance of certain birds, and by the flowering of particular plants. An attention to theſe phenomena is now ſuperſeded by our ſimple and accurate chronology; yet as a ſubject of curioſity, it ought not to be wholly diſregarded. With this view one of the diſciples of Linnæus conſtructed, at Upſal in Sweden, what he terms a *Calendar of Flora*. We have ſelected from it the circumſtances relating to birds, and have joined the obſervations made at Stratton in Norfolk, by Mr. Stillingfleet. Upſal is in latitude 59° 51′, Stratton in 52° 45′.

	At Stratton.		At Upsal.
The Wood Lark sings	February	4	March 20.
Rooks begin to pair -	———	12	
Geese begin to lay -	———	12	
The White Wagtail appears	———	12	April 13.
The Thrush sings - -	———	16	
The Chaffinch sings -	———	16	
Partridges begin to pair	———	22	
Rooks begin to build	March	2	
The Thrush sings	———	4	
The Ring Dove cooes	———	5	Lapwing returns.
The Swallow returns	April	6	Wild Duck returns.
The Nightingale sings	———	9	Swan & Land Rail re.
The Bittern makes a noise	———	14	Kestrel returns.
The Red-start returns	———	16	Turkey sits.
The Cuckoo sings -	———	17	May 12.
The Black cap sings	———	28	
The White-throat sings	———	28	May 5, the Stare retu.
The Goat-sucker heard	June	5	May 9, Swallow and Stork return.
The Nightingale sings	———	15	
The Rooks cease to resort in the evening to their nest-trees	about	21	May 15. Nightingale returns.
Young Partridges seen	July	18	July 15, the Cuckoo silent.
Hens moult - -	———	20	
The Ring Dove cooes	———	31	
The Nuthatch chatters	August	7	Birds of passage prepare for their departure.
Rooks visit their nest trees in the evening, without roosting	———	12	
The Stone Curlew whistles at night	———	14	
The Goatsucker makes a noise in the evening	———	15	
Rooks roost on their nest trees	———	17	
The Goatsucker no longer heard	———	17	
The Robin sings -	———	26	
The Chaffinch chirps	September	16	
Swallows gone - -	———	21	September 17, War tail departs.
The Wood-lark sings	———	25	

	At Stratton.	
The Fieldſare appears - -	September	25
The Blackbird ſings - -	———	27
The Thruſh ſings - -	———	29
The Royſton-Crow returns -	October	2
The Blackbird ſings - -	———	7
The Wood-lark ſings - -	———	10
The Ring Dove cooes -	———	10
The Wild Geeſe retire from the fens	———	16
The Woodcock returns -	———	22
The Sky-lark - - -	———	24

The following TABLES which we have extracted from Mr. WHITE's Natural Hiſtory of Selborne, exhibit a ſynoptic view of our ſinging birds, and illuſtrate the climate of the South of England.

LIST of the SUMMER BIRDS OF PASSAGE, ranged in the order of their appearance in the neighbourhood of Selborne in Hampſhire.

1	The Wryneck - -	The middle of March: harſh note.
2	The Smalleſt Willow-wren	March 23: chirps till September.
3	The Swallows and Martins	April 13.
4	The Black cap - -	April 13: ſweet wild note.
5	The Nightingale -	Beginning of April.
6	The Cuckoo - -	Middle of April.
7	The Middle Willow-wren	Middle of April: ſweet plaintive note.
8	The White-throat	Middle of April: mean note: ſings till September.
9	The Red-ſtart - -	Middle of April: more agreeable ſong.
10	The Stone Curlew	End of March: loud nocturnal whiſtle.
11	The Turtle Dove -	
12	The Graſhopper-lark	Middle of April: a ſmall ſibilous note till the end of July.
13	The Swift - -	About April 27.
14	The Leſs Reed-ſparrow	
15	The Land-rail -	
16	The Largeſt Willow-wren	End of April.
17	The Goatſucker -	Beginning of May.
18	The Flycatcher -	May 12.

A List of the Winter Birds of Passage.

1	The Ring-ouzel -	Michaelmas week and again about the 14th March.
2	The Redwing - -	About Old Michaelmas.
3	The Fieldfare -	
4	The Royston-crow -	
5	The Woodcock -	Appears about Old Michaelmas.
6	The Snipe	
7	The Jack-snipe	
8	The Wood-pigeon	
9	The Wild Swan	
10	The Wild Goose	
11	The Wild Duck	
12	The Pochard	
13	The Wigeon	
14	The Teal	
15	The Crosbill	Appear only occasionally.
16	The Grosbeak	
17	The Silktail	

A List of Birds that continue their Song till after Midsummer.

1	The Wood-lark -	From January to the end of autumn.
2	The Song-thrush - -	From February to August, and again in Autumn.
3	The Wren and Red-breast.	The whole year except in hard frost.
4	The Hedge-sparrow	Early in February till 10th July.
5	The Yellow-bunting	Early in February till 21st Aug.
6	The Sky-lark - -	From February to October.
7	The Swallow -	From April to September.
8	The Black-cap - -	Beginning of April till 13th July.
9	The Tit-lark -	Middle of April till 16th July.
10	The Black-bird -	February till 23 July, and again in autumn.
11	The White-throat -	April till 23d July.
12	The Gold-finch -	April till 16th September.
13	The Green-finch -	Till July and 2d August.
14	The Common Linnet	Whistles till August, and resumes its note in October.

A List of Soft-billed Insectivorous Birds which remain with us the whole Winter.

1 The Red-breast.
2 The Wren.
3 The Wagtails.
4 The Wheat-ear,
5 The Win-chat.
6 The Stone-chat.
7 The Gold-crested Wren.

A List of Birds that Sing while on Wing.

1 The Sky-lark -	Rising suspended, and falling.
2 The Tit-lark -	Descending; sitting on trees; and walking on the ground.
3 The Wood lark -	Suspended. Sometimes whole summer nights.
4 The White-throat	With jerks and gesticulations.
5 The Black-bird -	Sometimes from bush to bush.
6 The Swallow -	In soft sunny weather.
7 The Wren -	Sometimes from bush to bush.

Birds that Sing in the Night.

1 The Nightingale -	In Shady coverts.
2 The Wood-lark -	Suspended in Mid-air.
3 Less Reed-Sparrow	Among reeds and willows.

The following Table is borrowed from the Philosophical Transactions for 1773. Every person will not perhaps form the same opinion with Mr. *Barrington* on a subject so disputable as the comparative merits of Singing Birds. We insert it however for its curiosity.

A TABLE of the COMPARITIVE MERIT of the BRITISH SINGING BIRDS.

N. B. 20 is the point of absolute perfection.

	Mellowness of tone	Sprightly notes	Plaintive notes.	Compass.	Execution.
The Nightingale -	19	14	19	19	19
The Sky lark - -	4	19	4	18	18
The Wood-lark -	18	4	17	12	8
The Tit-lark - -	12	12	12	12	12
The Linnet -	12	16	12	16	18
The Gold finch -	4	19	4	12	12
The Chaf-finch -	4	12	4	8	8
The Green-finch -	4	4	4	4	6
The Hedge-sparrow	6	0	6	4	4
The Siskin - -	2	4	0	4	4
The Red poll - -	0	4	0	4	4
The Thrush - -	4	4	4	4	4
The Black-bird -	4	4	0	2	2
The Robin - -	6	16	12	12	12
The Wren -	0	12	0	4	4
The Reed-sparrow	0	4	0	2	2
The Black-cap -	14	12	12	14	14

END OF THE SEVENTH VOLUME.

Printed in the United States
By Bookmasters